T0341058

Computer-Aided Inspection Planning

Theory and Practice

Computer-Aided Inspection Planning

Theory and Practice

Abdulrahman Al-Ahmari
Emad Abouel Nasr
Osama Abdulhameed

CRC Press is an imprint of the
Taylor & Francis Group, an **informa** business

CRC Press
Taylor & Francis Group
6000 Broken Sound Parkway NW, Suite 300
Boca Raton, FL 33487-2742

Printed on acid-free paper
Version Date: 20161020

International Standard Book Number-13: 978-1-4987-3624-4 (Hardback)

Library of Congress Cataloging-in-Publication Data

Names: Al-Ahmari, Abdulrahman M., 1968- author. | Nasr, Emad Abouel, author. | Abdulhameed, Osama, author.
Title: Computer aided inspection planning : theory and practice / Abdulrahman Al-Ahmari, Emad Abouel Nasr, and Osama Abdulhameed.
Description: Boca Raton : Taylor & Francis, CRC Press, 2017. | Includes bibliographical references.
Identifiers: LCCN 2016026227 | ISBN 9781498736244 (hardback : alk. paper)
Subjects: LCSH: Engineering inspection--Data processing. | Computer integrated manufacturing systems. | Computer-aided engineering.
Classification: LCC TS156.2 .A425 2017 | DDC 620.0028/5--dc23
LC record available at https://lccn.loc.gov/2016026227

Visit the Taylor & Francis Web site at
http://www.taylorandfrancis.com

and the CRC Press Web site at
http://www.crcpress.com

Printed and bound in the United States of America by Sheridan

Contents

Description of This Book

Inspection process is one of the most important steps in many industries, including manufacturing, which ensures high-quality products and customer satisfaction. The process verifies whether the manufactured part lies within the tolerance of the design specifications. Manual inspection may not provide the desired process accuracy due to human interaction and involvement in the process. There are many factors such as fatigue, workspace, lack of concentration of the operator, and accurate inspection steps that degrade the performance of the process with time. In many cases, manual inspection is not feasible because the part's size or high production rate. Thus, automated inspection provides the necessary solution to many problems associated with the manual inspection. The automation of the inspection process will increase the productivity. This book introduces a new methodology, providing methods for its implementation, and also describes the supporting technologies for automated inspection planning based on computer-aided design (CAD) models. It also provides an efficient link for automated operation based on coordinate measuring machine (CMM) giving details of its implementation. The link's output is a DMIS code programming file based on the inspection planning table that is executed on CMM.

Approach

This book offers insights into the methods and techniques that enable implementing inspection planning by incorporating advanced methodologies and technologies in an integrated approach. It includes advanced topics such as feature-based design and automated inspection. This book is a collection of the latest methods and technologies. It will be structured in such a way that it will be suitable for a variety of courses in design, inspection, and manufacturing. Most books developed in the inspection area are very theoretical (Anis Limaiem, etc.), although this book is designed in such as way to address more practical issues related to design and inspection. This book includes a discussion of the theoretical topics, but the focus is mainly on applications and implementations contexts.

This book is the result of an extensive research and development in this area. The proposed methodology has been implemented, tested, and validated.

Target Readership

- Institute of Industrial Engineers (IIE)
- Society of Manufacture Engineering (SME)
- INFORMS
- Engineering Management Society
- American society of Mechanical Engineers (ASME)

Authors

Abdulrahman Al-Ahmari is the dean of Advanced Manufacturing Institute, executive director of Center of Excellence for Research in Engineering Materials (CEREM), and supervisor of Princess Fatimah Alnijris's Research Chair for Advanced Manufacturing Technology. He earned his PhD in manufacturing systems engineering in 1998 from the University of Sheffield, UK. His research interests are in analysis and design of manufacturing systems, computer-integrated manufacturing (CIM), optimization of manufacturing operations, applications of simulation optimization, flexible manufacturing system (FMS), and cellular manufacturing systems.

Emad Abouel Nasr is an associate professor in the Industrial Engineering Department, College of Engineering, King Saud University, Saudi Arabia, and Mechanical Engineering Department, Faculty of Engineering, Helwan University, Egypt. He earned his PhD in industrial engineering from the University of Houston, Texas, in 2005. His current research focuses on CAD, CAM, rapid prototyping, advanced manufacturing systems, supply chain management, and collaborative engineering.

Osama Abdulhameed is a PhD candidate of industrial engineering at King Saud University, Riyadh. He received his master's degree in industrial engineering from King Saud University in 2013. His research activities include advanced manufacturing, CAD, CAM, with additive manufacturing as his main focus and interest.

Chapter 1

Computer-Based Design and Features

1.1 Introduction

Manufacturing industries have been directing their efforts to decrease the cost of their products [1]. This has become necessary because of the rapidly increasing demands, insistence on higher quality, better performance, and on-time delivery, in addition to reasonable cost. These parameters are very important in order to thrive and sustain in a highly competitive global market. According to Nasr and Kamrani [2], the quality of the product design (i.e., how well the product has been designed) has been one of the crucial factors in establishing the commercial success as well as a societal value of the product. Therefore, the product design can be considered as one of the most significant operations in the design life cycle. Moreover, it is worth mentioning here that a substantial fraction of the total cost of any product depends on its design procedures or techniques [3]. The importance of design in manufacturing can further be emphasized by the fact that approximately 70% of the manufacturing costs of the product

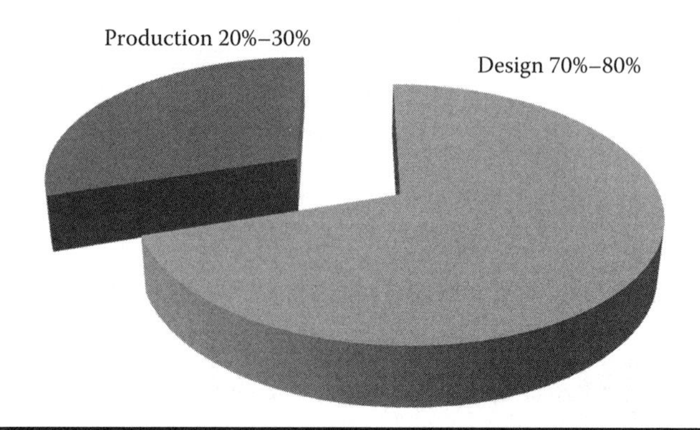

Figure 1.1 Contribution of the design and production costs to the total cost of the final product.

depend on the design decisions while production decisions only contribute to about 20% [4] (Figure 1.1).

Design can be defined as a process of developing a system, component, or process to meet the customer requirements [5]. It is, in fact, a decision-making process that involves the implementation of basic sciences, mathematics, and engineering technologies to transform resources optimally to achieve the desired goal [6]. The design process is actually one of the primary components in the sequential cycle, which include the design department, process planning department, and the manufacturing department. It has widely been considered as one of the most important steps in the development of any product [7]. This is due to the fact that a poorly designed product would always result in a failed end product. Walton [8] identified a number of reasons responsible for poor engineering designs:

- Inappropriate and unnecessary assumptions
- Lack of sufficient information or the understanding of the problem
- Erroneous design specifications
- Incorrect design calculations
- Inadequate data collection and experimentation
- Incorrect drawings, etc.

Manufacturing, on the other hand, is a process of converting raw materials and design information into finished components to the satisfaction of customer requirements [9]. In this highly competitive market, the manufacturing industries have been facing many challenges in terms of cost and time reduction, quality and flexibility improvement, etc. Therefore, there are a number of factors that should be considered by the manufacturing industries to survive in the globally competitive market [1].

- Reduction in inventory
- Minimization of waste
- Enhancement of quality
- Higher flexibility in terms of
  - Product changes
  - Production changes
  - Process change
  - Equipment change
  - Change of personnel

Design and manufacturing are the two primary driving forces to accomplish an efficient engineering process. A product design that cannot be successfully achieved through the available manufacturing processes is a poor design. Similarly, the manufacturing processes are ineffective without a reasonable design and plan [10]. Conventionally, design and manufacturing have been considered as two separate entities in a product development cycle. Since they have been carried out by two sets of people, there is no communication between the two groups, that is, no information flow [11]. The detailed design in the form of annotated engineering drawing is passed on to the manufacturing personnel to get the end product. Most often, it takes a number of runs between the two groups until they reach a satisfactory conclusion [12]. This results in a slow and costly process. Thus, it is very important for the manufacturing industries to implement novel techniques in different phases of the product development cycle. Hence, to accomplish the task of efficient and cost-effective production, Boyer [13] emphasized on the importance of integration between the design and manufacturing processes. The seamless integration between the different stages provides a provision of real-time response to changes in design, setup planning, production scheduling, etc. [9]. In fact, the primary objective of the integration of computer-aided design and the computer-aided manufacturing is to assist the design, modification, analysis, and manufacture of parts automatically and efficiently within the specified time [2].

1.2 Computer-Aided Design

Computer-aided design (CAD) can be defined as a design process involving the generation of digital (computer) models using various geometrical parameters such as angles, distances, and coordinates [14]. It can also be defined as the technology to produce technical drawings and plans to finally manufacture products for various industries such as aerospace, automobiles, medical, and oil and gas pipelines [2]. CAD drawings are helpful because they provide substantial information in the form of technical details of the product, dimensions, materials, and procedures. The working of the CAD systems is based on the generation and storage of drawings electronically, which can be viewed, printed, or programmed directly into the automated manufacturing systems. It enables the designers to view objects under a wide variety of representations and to test these objects through simulations [14]. One of the main benefits of the CAD over traditional methods is that the CAD models can be modified or manipulated by varying geometrical parameters. Moreover, the design can be tested or verified through simulation in CAD systems. The CAD also promotes the flow of the design process to the manufacturing process through numerical control (NC) technologies. A CAD comprises of three basic elements:

- Geometrical modeling and computer graphics: Generation and visualization of models.
- Analysis and optimization tool: Predicting the behavior of the model with all constraints and boundary conditions.
- Drafting and documentation.

Meanwhile, the modeling with CAD systems provides several advantages over traditional drafting methods, which uses rulers, squares, and compasses [14]:

- Designs can be altered without erasing and redrawing.
- CAD systems also offer "zoom" features whereby designers can magnify specific elements of a model to carry out visual inspection.
- CAD models are three-dimensional (3D) and therefore can be rotated on any axis aiding the designers to experience the complete object.
- CAD systems also provide section views to look into the internal shape of a part and illustrate the spatial relationships between various entities of the part.

It should be pointed out here that the CAD does not only generate geometrical shapes, but also represents specific functions on individual shapes, thus providing physical properties.

1.3 Computer-Aided Manufacturing

The use of computer-based software tools to support engineers and machinists for the manufacturing of product components is termed as the computer-aided manufacturing (CAM). The primary objective of the CAM systems is to generate necessary instructions for operating manufacturing systems [15]. In fact, the CAM provides geometrical design data to manage automated machines. The development of the high-performance CAM systems requires the following basic information [15]:

- Decision tables
- Generative algorithms and engineering drawings
- Adequate data about the manufacturing technologies

CAM can also be defined as a programming tool to manufacture physical models through CAD programs. According to Techopedia definition [16], CAM is an application technology that uses computer software and machinery to assist and automate the manufacturing processes. The working of the CAM systems is based on the encoding of the geometrical data using computer numerical control (CNC) or direct numerical control (DNC) systems. Conventionally, the CAM has been considered as a NC programming tool, wherein the two-dimensional (2D) or three-dimensional (3D) models created with CAD software are used to generate G-code, which drive CNC machine tools [17] (see Figure 1.2). The primary benefits of the CAM in production systems are manifold:

- Faster production process
- Parts with more precise dimensions and material consistency
- Minimization of waste
- Reduction of energy consumption, etc.

CAM reduces waste, energy, and enhances manufacturing and production efficiencies through increased production speeds, raw material consistency, and precise tooling accuracy.

The performance of the CAM systems can be improved by collecting a good database about the manufacturing technologies [15]. The application of the modern CAM

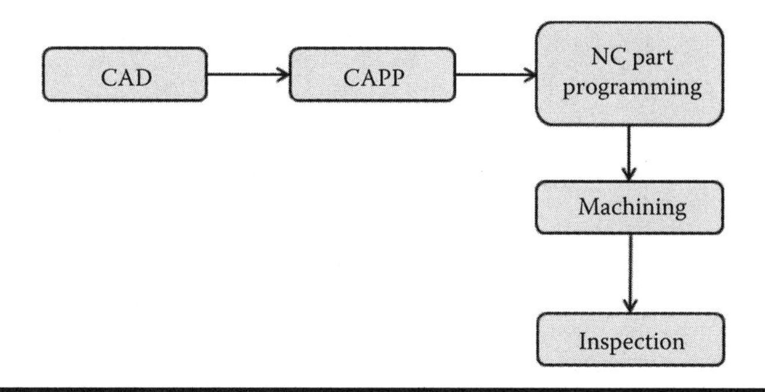

Figure 1.2 Different phases of CAM. (Adapted from A. Dwivedi and A. Dwivedi, *International Journal of Innovative Technology and Exploring Engineering (IJITEE)* 2013;3(3):174–181.)

solutions can range from discrete systems to multi-CAD 3D integration. CAM can be related with CAD in order to achieve improved and streamlined manufacturing, efficient design, and superior machinery automation [16]. Since both the CAD and CAM utilize computer-based methods, therefore, it is possible to integrate both design and manufacturing processes. CAD and CAM, which are not fully integrated, require specialists to translate the output of design into input information for the CAM systems. The integration of the CAD and CAM systems is based on the following prerequisites [15]:

- Knowledge of the designer's objective from the results of the design
- Understanding and application of manufacturing technologies

1.4 CAD and CAM Integration

The processes of design and manufacture are two conceptually independent operations. However, the design process has to be carried out in synchronization with the knowledge of the nature of the production process. This is due to the fact that the designer must have the prior knowledge of the properties of the machining materials, various machining techniques, and the production rate. Therefore, the integration between design and manufacture can provide potential benefits of both CAD and CAM systems [14]. Currently, the trend in the market requires companies to be competitive enough in terms of low cost, high quality, and lesser delivery times. These requirements for survival can effectively be achieved through the integration of CAD and CAM systems as shown in Figure 1.3. The proper integration of the CAD and CAM systems can help to survive the increasingly stringent demands of the productivity and quality in the design and production [18].

The primary steps in the CAD/CAM integration can be explained as follows [20,21]:

- Product design: The design has to be carried out with prior knowledge of product applications and functions. This can be achieved through various stress and strain analysis using appropriate CAD software. The output of this step is an appropriate design in terms of optimized shape and size.

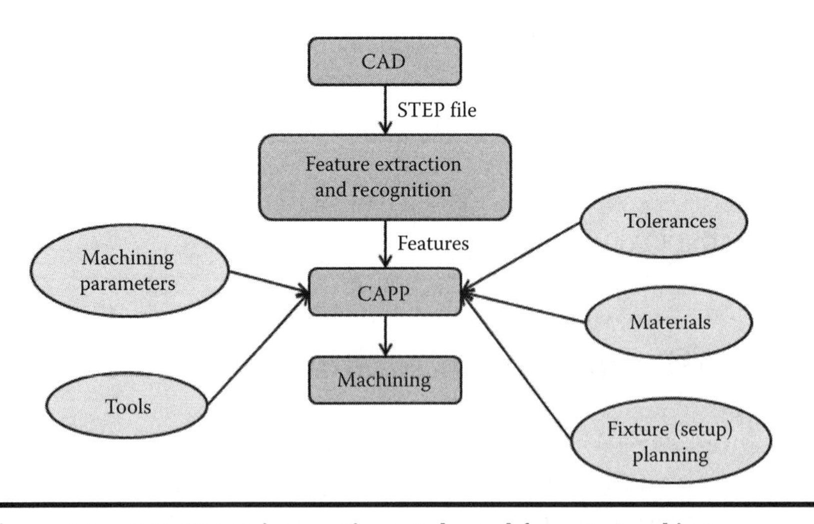

Figure 1.3 CAD/CAM integration. (Adapted from J. Saaski, T. Salonen, and J. Paro, *Integration of CAD, CAM and NC with Step-NC*, VTT Information Service, 2005, Espoo, Finland.)

- Drafting and documentation: Once the designing of the product is finished, the assembly drawings and the part drawing are prepared using CAD software. These drawings are to be used as the blueprint during the manufacturing of the product on the shop floor.
- Production planning and scheduling: This phase defines the CAM phase of the CAD/CAM integration. This phase of the production planning and scheduling includes management of manufacturing resources such as tools, materials, fixture set up, machining parameters, and tolerances.
- Manufacturing: In this phase, the machining instruction generated in the earlier phases is fed to the CNC machines. The generated program provides appropriate instructions to perform the manufacturing of the product according to the prescribed dimensions.

1.5 Role of CAD/CAM in Manufacturing

Since 1970, there has been a rapid growth in the use of CAD/CAM technologies, primarily due to the development of high-performance computer systems [22]. The inventions of silicon chips and microprocessors have resulted in the more cost-effective computers afforded by even small companies [14]. These developments have expanded the horizon of CAD/CAM technologies from large-scale industries to setups of all sizes. In fact, the CAD/CAM has extensively been used by aerospace, automotive, medical industries, in addition to, companies involved in the production of consumer electronics, electronic components, and molded plastics. There has been a great need toward the development of a single CAD–CAM standard, so that information in different data systems can be exchanged without delays and unnecessary changes [14].

The integration of the CAD and CAM systems have overcome most of the limitations of the conventional machining in terms of cost, ease of use, and speed. Moreover, the CAD/

CAM integration provides the industrial personnel greater control over the production processes. The CAD/CAM integration promotes streamlined flow of information between the various departments such as design, manufacturing, and inspection. [23]. The effective implementation of CAD/CAM systems offers companies several benefits including reduced design cost, lesser machining time and overall cycle time, and smooth information flow [24]. The CAD/CAM integration helps manufacturing sectors through the better tool design and optimization of the manufacturing processes. The powerful CAD/CAM systems, which can create a virtual manufacturing environment, can avoid many uncertainties such as time delay, rework, and defective parts through simulation [25]. With time, the CAD/CAM systems have evolved to include many functions in manufacturing, such as material requirements planning, production scheduling, computer production monitoring, and computer process control [2]. In manufacturing industries, the ideal CAD/CAM system is the one that ensures an automatic streamlined flow of the design specification from the CAD database through process plan to the CNC on the shop floor [26]. According to Nasr and Kamrani [2], the features provide the basis to link the CAD with the downstream applications as shown in Figure 1.4.

1.6 Feature-Based Technologies

The 3D CAD models can be used for visualization saving much effort in prototype fabrication, thus making it easy to integrate with manufacturing functions. The geometric data for the design (or the CAD models) can be represented using a number of feature representation methods such as wireframe representation, boundary representation (Brep), or constructive solid geometry (CGS). These feature representation methods have been detailed in the subsequent chapters. Once the geometric model is constructed, this geometric data has to be transferred

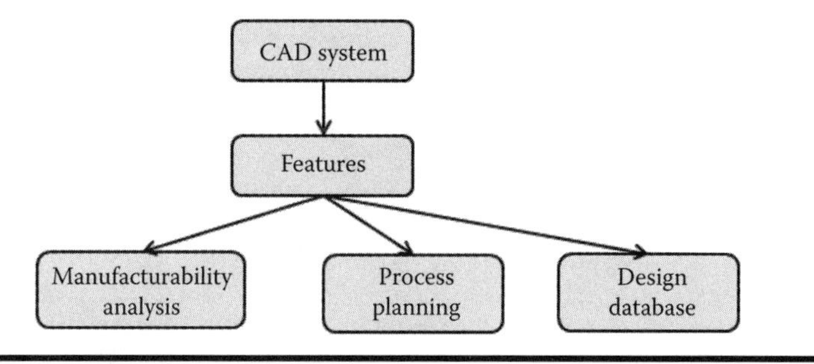

Figure 1.4 Features acting as a link between design and downstream applications.

into a format that can be used to generate the required information for the manufacturing processes. This process of conversion of geometric data is called feature recognition or feature extraction [2]. There have been a number of approaches such as graph matching, syntactic recognition, volume decomposition, and rule-based algorithms, which can be used for the feature recognition. A systematic flow of the information through various phases of CAD/CAM systems can be seen in Figure 1.5.

1.6.1 Types of Features

The feature-based approach can be defined as a design representation method in which the design is expressed in terms of high level definition to be used in downstream manufacturing activities such as process planning [28]. The features can be classified based on their applications as shown in Figure 1.6. For example, in the feature-based design, holes, slots, pockets, and steps represent manufacturing features as compared to traditional CAD where design is either in terms of 2D entities (lines, arcs, or circles) or 3D entities (wireframe, surfaces, or solids). The feature information has a greater significance because it helps the process planner to determine the machining tools and manufacturing processes required to machine the designed objects.

Geometric features can be classified either as form features or primitive features based on their functions [29,30]. Form features are the specific shapes or configurations such as holes, slots, and

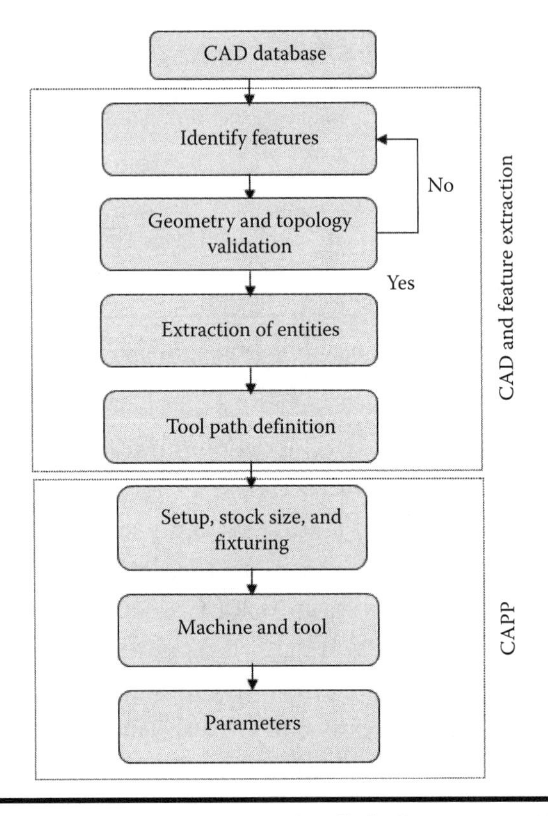

Figure 1.5 Features as interconnecting links between various phases of CAD/CAM. (Adapted from S. Somashekar and W. Michael, *Computers in Industry* 1995;26(1):1–21.)

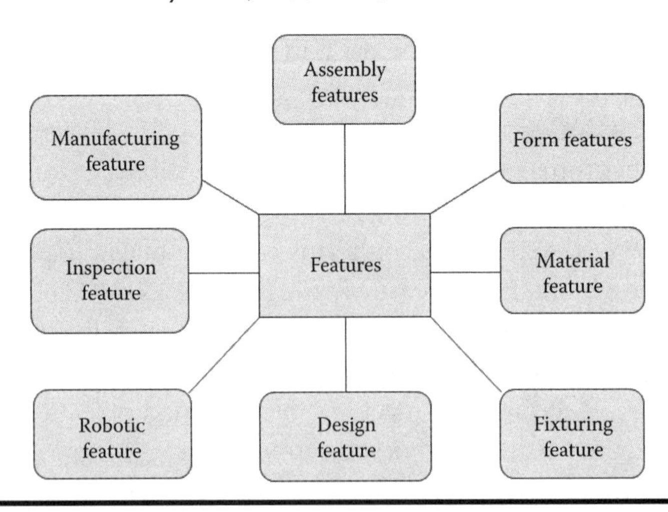

Figure 1.6 Types of features.

chamfers, which are produced on surfaces, edges, or corners of a part. Their primary purpose is to accomplish a specific task or alter the appearance of the part. On the contrary, a primitive feature can be defined as a basic geometric entity of a part, such as surfaces, edges, and vertices. In fact, form features are built on the top of primitive features. They are either added to or subtracted from primitive features to achieve a given design or manufacturing functions. Similarly, design features are defined as a set of geometric entities that represent particular shapes, patterns and possess certain functions or embedded information [31]. Moreover, manufacturing features can be defined as a section of the workpiece that can be created using metal removal processes [31]. Machining (or manufacturing) features can also be defined both as surface features as well as volumetric features [32]. When defined as surfaces, machining features are a group of faces that are to be created using a given machining operation. A machining feature usually corresponds to the volume of material that can be removed by a machining operation. Generally, geometry and tolerance information that can correspond with the design attributes of the part and parameterized the manufacturing operations is associated with manufacturing features [33]. The material features define material composition and treatment condition [34]. Moreover, an assembly feature can be defined as an association between two form features which exists in different parts, that is, geometry that belongs to different parts [35]. Actually, assembly features convert the mutual constraints on mating feature's shape, dimensions, position, and orientation. It can further be defined as a grouping of various features that define assembly relations such as mating conditions, position, orientation, and kinematic relations [36].

1.7 Summary

This chapter provides an overview of computer-aided design and manufacturing (CAD/CAM) as well as an overview of CAD/

CAM integration. It also explains the role of CAD/CAM systems in the manufacturing facility and provides a discussion about feature-based technologies and the different types of features.

QUESTIONS

1. What is the difference between the design and manufacturing?
2. Why design is considered highly significant in the manufacturing sector?
3. What do you mean by poor engineering design?
4. What are the factors responsible for poor engineering design?
5. List down the factors that are critical for the existence of manufacturing industries.
6. Define CAD and CAM.
7. Write down the basic elements of CAD.
8. What are the several benefits of CAD systems over traditional drafting methods?
9. What are the inputs required for the development of CAM system?
10. Discuss the benefits of CAM in production systems.
11. What are prerequisites for the integration of CAD and CAM systems?
12. How manufacturing industries can be benefited with CAD/CAM integration?
13. What are the primary steps required for CAD/CAM integration?
14. Explain the role of CAD/CAM in manufacturing.
15. What are the different methods for feature representation?
16. Write down the different approaches that can be used to perform feature recognition.
17. How do features act as the interconnecting link between different phases of CAD/CAM?
18. What is the difference between form features and primitive features?

19. Explain the following terms:
    a. Design features
    b. Manufacturing features
    c. Assembly features

References

1. D. E. Culler and W. Burd, A framework for extending computer aided process planning to include business activities and computer aided design and manufacturing (CAD/CAM) data retrieval, *Robotics and Computer-Integrated Manufacturing* 2007;23(3):339–350.
2. E. A. Nasr and A. K. Kamrani, *Computer-Based Design and Manufacturing: An Information-Based Approach*. Berlin: Springer, 2007.
3. M. P. Groover, *Automation, Production Systems, and Computer-Integrated Manufacturing*. Englewood Cliffs, NJ: Prentice-Hall, 2001.
4. T.-C. Chang, R. A. Wysk, and H.-P. Wang, *Computer-Aided Manufacturing*, Second Edition, Upper Saddle River, NJ: Prentice-Hall, United States, 1998, 596–598.
5. T.-C. Kuo, S. H. Huang, and H.-C. Zhang, Design for manufacture and design for "X": Concepts, applications, and perspectives, *Computers and Industrial Engineering* 2001;41(3):241–260.
6. Accreditation, 1988, Accreditation Board for Engineering and Technology (ABET), Inc. Annual Report for the year ending, New York.
7. Y. Haik and T. Shahin, *Engineering Design Process*, 2nd ed. New York: Cengage Learning, University of North Carolina—Greensboro, Kings College London, UK, May 14, 2010.
8. J. Walton, *Engineering Design: From Art to Practice*, New York: West Publishing Company, 1991.
9. A. Y. C. Nee, S. K. Ong, G. Chryssolouris, and D. Mourtzis, Augmented reality applications in design and manufacturing, *CIRP Annals—Manufacturing Technology* 2012;61(2):657–679.
10. D. D. Bedworth, M. R. Henderson, and P. M. Wolfe, *Computer Integrated Design and Manufacturing*, New York: McGraw Hill, 1991.

11. R. Lionel, S. Otto, and P. Henri, Process planning as an integration of knowledge in the detailed design phase, *International Journal of Computer Integrated Manufacturing* 2003;16(1):25–37.
12. D. Pham and C. Gologu, A computer aided process planning system for concurrent engineering, *Process and Instrumentation Mechanical Engineers, Part B: Journal of Engineering Manufacture* 2002;215(1):1177–1131.
13. K. K. Boyer, Evolutionary patterns of flexible automation and performance: A longitudinal study, *Management Science* 1999;45(6):824–842.
14. R. K. Shukla and D. B. Deshmukh, A review on role of CAD/CAM in designing for skill development, *International Journal of Research in Engineering, Science and Technologies (IJRESTs)* 2015;1(2):4–7.
15. H. Yoshikawa and K. Uehar, Design theory for CAD/CAM integration, *CIRP Annals—Manufacturing Technology* 1985;34(1):173–178.
16. Y. Asiedu and P. Gu, Product life cycle cost analysis: State of the art review, *International Journal of Production Research* 1998;36(4):883–908.
17. A. Dwivedi and A. Dwivedi, Role of computer and automation in design and manufacturing for mechanical and textile industries: CAD/CAM, *International Journal of Innovative Technology and Exploring Engineering (IJITEE)* 2013;3(3):174–181.
18. J. Majerik and J. Jambor, Computer aided design and manufacturing evaluation of milling cutter when high speed machining of hardened steels, *Procedia Engineering* 2015;100:450–459.
19. J. Saaski, T. Salonen, and J. Paro, *Integration of CAD, CAM and NC with Step-NC*, VTT Information Service 2005, Espoo, Finland, ISSN: 1459–7683.
20. M. P. Groover and E. W. Zimmers, *CAD/CAM: Computer Aided Design and Manufacturing*, Upper Saddle River, NJ: Prentice-Hall, 1997.
21. H. Yoshikawa and K. Uehara, Design theory for CAD/CAM integration, *CIRP Annals—Manufacturing Technology* 1985;34(1):173–178.
22. A. A. Adekunle, S. B. Adejuyigbe, and O. Faluyi, Material selection for computer aided design software for crankshaft design, *International Journal of Scientific and Research Publications* 2014;4(8).
23. A. Kamrani, E. A. Nasr, A. Al-Ahmari, O. Abdulhameed, and S. H. Mian, Feature-based design approach for integrated CAD and computer-aided inspection planning, *International Journal*

of Advanced Manufacturing Technology 2015;76:2159–2183. DOI: 10.1007/s00170-014-6396-0.

24. F. Soliman, S. Clegg, and T. Tantoush, Critical success factors for integration of CAD/CAM systems with ERP systems, *International Journal of Operations and Production Management* 2001;21(5/6):609–629.
25. P. Sharma, K. Pathak, and B. K. Sharma, Role of CAD/CAM in designing, developing and manufacturing of new products, *International Journal of Research in Engineering and Technology* 2014;3(6):146–149.
26. K. Lee, *Principles of CAD/Cam/CAE Systems.* Reading, MA: Addison Wesley, 1999.
27. S. Somashekar and W. Michael, An overview of automatic feature recognition techniques for computer-aided process planning, *Computers in Industry* 1995;26(1):1–21.
28. M. A. Chamberlian, Protrusion-features handling in design and manufacturing planning, *Computer-Aided Design Journal* 1993;25(1):19–28.
29. N. Wang and T. M. Ozsoy, Representation of assemblies for automatic tolerance chain generation, *Engineering with Computers* 1990;6:121–126.
30. N. Wang and T. M. Ozsoy, A scheme to represent features, dimensions and tolerances in geometric modeling, *Journal of Manufacturing Systems* 1991;I0(3):233–240.
31. T. M. M. Shahin, Feature-based design—An overview, *Computer-Aided Design and Applications* 2008;5(5):639–653.
32. A. Singh, *Manufacturing Feature Recognition from Solid Models*, Department of Mechanical & Industrial Engineering, Thapar Institute of Engineering and Technology, June 2002.
33. S. K. Gupta, W. C. Regli, and D. S. Nau, Manufacturing feature instances: Which ones to recognize? In *3rd Symp. on Solid Modeling and Applications*, ed. C. Hoffmann and J. Rossignac, May 7–19, 1995. Salt Lake City, UT: ACM Press, pp. 141–152.
34. J. J. Shah, Assessment of features technology, *Computer-Aided Design*, 1991;23(5):331–343.
35. J. J. Shah and M. T. Rogers, Assembly modeling as an extension of feature-based design, *Research in Engineering Design* 1993;5:218–237.
36. C. Hoffmann and R. Joan-Arinyo, Parametric modeling, *Handbook of Computer Aided Geometric Design*, eds. G. Farin, J. Hoschek, and M.-S. Kim, Amsterdam: Elsevier, 2002. ch. 21, pp. 519–541.

Chapter 2

Methodologies of Feature Representations

2.1 Feature Definitions

A feature can be defined either as a section of a part, which possesses some manufacturing specifications [1], or a geometric shape, which can be utilized in computer-aided design (CAD) [2]. In fact, a functional entity (object, shape, or process), which is meaningful in a certain domain (for example, a fastener in assembly, a shaft in design, or a groove in machining), [3] represents a feature. Broadly speaking, a feature can also be defined as an entity that represents a general shape such as holes, pockets, slots, ribs, or bosses, achieved through a manufacturing operation on the raw stock [4,5]. The word "features" indicates several meanings in different contexts [4]. For example, in design, it refers to a web, or an aerofoil section, while in manufacturing, it refers to the slots, holes, and pockets, while in inspection it is used as a datum or reference on the part. The features can be classified as shape features, manufacturing features, assembly features, and geometric features depending on their application requirements [6–9]. Moreover, the features can be additive such as bosses and webs, as well as subtractive such as holes and slots. The different features can be categorized as follows [10,11]:

- Form features represent portions of nominal geometry
- Tolerance features provide deviations from nominal form/size/location
- Assembly features can be considered as a collection of various features types to define assembly relations, such as mating conditions, part relative position and orientation, various kinds of fits, and kinematic relations
- Functional features represent sets of features related to specific function. It includes design intent, nongeometric parameters related to function, performance, etc.
- Material features comprise of material composition, treatment, condition, etc.

The features can also be defined as explicit features where all the details of the features are fully defined and implicit features where only sufficient information is provided to define the features [12]. According to Shah and Rogers, [13], any entity that possesses the following characteristics can be recognized as a feature:

- A feature is a physical constituent of a part.
- A feature is mappable to a generic shape.
- A feature has engineering significance.
- A feature has predictable properties.

A feature represents the engineering significance of the geometry of a part [14]. For example, a flat surface, a hole, and a chamfer can be considered as a feature. They are represented by geometric information, including a feature's shape, dimension, and nongeometric information such as form tolerances and surface finish. The part information is made up of a feature information and the relationships between features, such as dimension, position, and orientation tolerances [15]. According to Amaitik [12], form features, tolerance features, and assembly features are all closely related to the geometry of parts, and hence collectively called as geometric features. Furthermore, the geometric features, according to their functions, can be categorized as form features and primitive features [16]. The purpose of the form features is to accomplish a given function or change the appearance of the part. Holes, slots, and chamfers represent the form features and they can be defined as the specific configurations produced on the surfaces, edges, or corners of the part. On the contrary, the primitive features can be defined as the basic entity of the part, such as surfaces, edges, and vertices or the geometric attribute of the part such as the center lines (axes) or center planes. In fact, the form features are created on the top of the primitive features. The form features are either added to or subtracted from the primitive features in order to attain certain design or manufacturing functions. The primary features are referenced while defining the dimensions and tolerances and specifying the mating features in the assembly representation. The features can further be classified as design features or machining features [17]. The design features are defined as the shapes controlling the part's function, its design objective, or the model construction methodology. On the contrary, machining features comprise the shapes that are associated with distinctive machining operations.

2.2 Features in Manufacturing

The implementation of feature-based modeling in manufacturing applications can associate design features with manufacturing process models. For example, the process model for a machining process would provide information regarding the process resources, such as machines, tools, fixtures, and auxiliary materials; process kinematics, such as tool access direction; process constraints, such as interference and spindle power; process parameters, such as feeds and speeds, and other information, such as time and cost [18]. It is the need of the hour to have a methodology or techniques that can integrate features, process models, and resource models efficiently.

2.2.1 Process Planning

Process planning can be defined as a procedure that involves the determination of the information required for manufacturing a given component. The automated process planning has been a key requirement for the integration of the design and manufacturing processes. The two primary approaches for automated process planning are variant process planning and generative process planning.

2.2.1.1 Variant Process Planning

The variant process planning (VPP) approach can also be called as a data retrieval method [19]. It involves retrieving an existing plan of a similar component and making the necessary modifications (if necessary) to prepare the plan for the new component. In fact, the process plan for a new component is generated by retrieving an existing plan for a similar component and making the necessary modifications for the new component. The features of a VPP can be discussed as follows [20]:

- Utilization of similarity among components to retrieve the existing process plans.
- A process plan that can be used by a family of components is called a standard process plan.
- Some modification of standard plan is always required to use it for the new component.
- It consists of two operational stages: preparatory and production stage.

The preparation stage involves coding of the existing components, part family formation and formation of family matrix, preparation of the standard process plan, among others. The similarity in design attributes and manufacturing methods are utilized for the formation of part families.

The production stage involves coding of the incoming component, the search routine to find the family to which the item belongs, or the retrieval of the standard process plan. In fact, it involves retrieving and modifying the process plan of master part of the family.

The various steps for VPP are as follows:

- Formation of the part families using group technology
- Development of a standard process plans
- Retrieval and modification of the standard plan for new components

- *Advantages of variant process planning approach*:
  - Reduced processing time and labor requirements
  - Utilization of standardized procedures
  - Reduced development and hardware cost and shorter development time

- *Disadvantages of variant process planning approach*:
  - Difficult to maintain consistency during modification
  - Difficult to achieve the combinations of attributes such as material, geometry, size, precision, quality, alternate processing sequence, and machine loading
  - The quality of the final process plan largely depends on the knowledge and experience of a process planner

2.2.1.2 Generative Process Planning

The generative process planning (GPP) approach involves the generation of a new process plan by means of decision logic, formulas, algorithms, and process knowledge. The primary objective is to convert a component from the raw material to the finished product. Therefore, it can be defined as a system that synthesizes process information in order to create a process plan for a new component automatically [19]. The generative process plan is made of two components:

- Geometry-based coding scheme
- Proportional knowledge in the form of decision logic and data

- *Advantages of generative process plan*:
  - They don't depend on group technology code numbers because it uses a decision tree to categorize parts into families.
  - Maintenance and updating of stored process plans is not required.
  - The use of process logic rules in its development ensures the quality of the process plan.
  - Lesser dependency on the expertise of the process planner.
  - Generate consistent process plane efficiently.

- *Disadvantages of generative process plan*:
  - The initial development of the system is difficult.
  - It requires major revisions (or modifications) with the introduction of new equipment or processing capabilities.

2.2.2 Assembly Planning

The assembling of individual components has been a key element in the manufacturing of products. This is due to the fact that most of the manufacturing products are made of assemblies of individual components. This leads to the need of an efficient assembly planning in the manufacturing sector. Generally, assembly planning can be divided into three phases [21]:

- *Selection of assembly method*: Identification of the suitable method depending on the type of assembly system to be used
- *Assembly sequence planning*: Generation of a sequence of assembly operations (placing each component in its final position in the assembly)
- *Assembly operations planning*: Description of the details of individual assembly steps, such as access directions, mating movements, and application of fasteners

The poor assembly planning leads to many problems in the production system such as poor quality, inefficiency, and high cost. Therefore, the concept of design for assembly approach (DFA) has been introduced. In this approach, the most economical assembly process is selected during design and the product is modified to the chosen method [22]. In the robotic assembly process, the operation planning involves the designing of a valid sequence of rigid motions, translations, and rotations that can bring a component into its final position in the assembly. Therefore, various assembly tools such as including grippers, jigs, fixtures, pallets, and component feeders have to be designed and assembly cells must be laid out while monitoring the physical restrictions of robot motions [23]. A feature model can be very helpful for assembly because it can record the physical arrangement of components in an assembly, as well as attribute information on the physical fit between linked components. These data are fundamental to carry out the necessary assembly operations and perform sequencing of these operations.

2.2.3 Inspection Planning

Six Sigma has been recognized as one of the most important techniques to ensure error free manufacturing [24]. The feature of this technique is that it ensures that defective parts are never produced. Therefore, part inspection has been vital not only to discard bad parts, but also to provide closed loop control on the quality of the products being produced. Meanwhile, the inspection planning is considered as one of the integral parts of manufacturing planning. The inspection plan should focus on the product characteristics, which significantly influences the performance of the product. It should also minimize the cost of inspection by reducing the inspection time. The inspection planning is affected by many issues such as product modeling issues, including representations of dimensions and tolerances, assembly relationships, functions and behaviors of physical configurations, etc. [25].

2.3 Geometric Modeling

Geometric modeling represents a collection of methods that help the scientists or engineers in the synthesis, representation, and analysis of shape and other geometric information [26]. The evolution of computer-based geometric modeling tools, which can be used to explain mechanical parts and assemblies, has been of great interest, especially in the engineering community. Broadly speaking, geometric modeling can be defined as the mathematical description of the geometry of the object [27]. Mortenson [28] defines geometric modeling as the technique to explain the shape of an object. Moreover, the geometric models play a significant role in the development of a computer-based environment and helps to integrate the various phases of the engineering design production cycle. In fact, the geometric modeling provides an analytical, mathematical, and abstract representation of an object in CAD [29]. There have been three different methods for representing the objects in geometric models (Figure 2.1): wireframe modeling, surface modeling, and solid modeling.

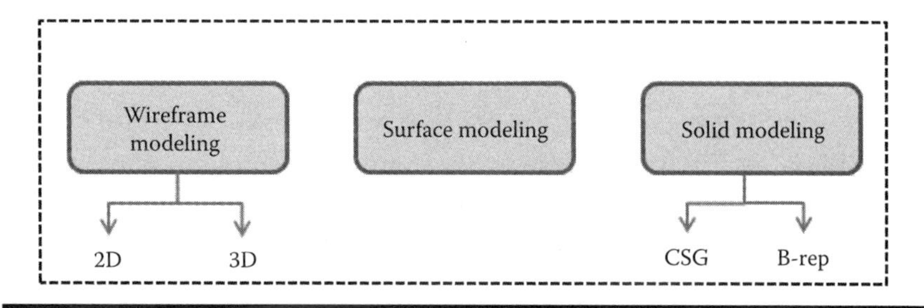

Figure 2.1 Geometric modeling methods. (Adapted from K. Holly, Ault, *Engineering Design Graphics Journal*, 1999;63(2):33–42.)

2.3.1 Wireframe Modeling

Wireframe modeling can be defined as the geometric modeling technique where the objects are represented using edges (line and curves) and vertices as shown in Figure 2.2. This suggests that the wireframe model does not have face

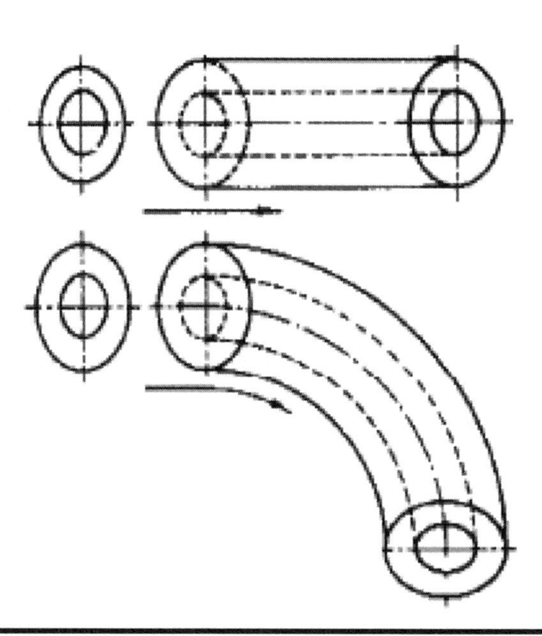

Figure 2.2 Representation of objects using wireframe modeling. (Adapted from J. Rooney and P. Steadman, Principles of Computer-Aided Design, London: UCL Press, 1997; O. Ostrowsky, Engineering Drawing with CAD Applications, Routledge, UK, 1989.)

information. Actually, the objects in the 2D wireframe consist of a collection of lines, arcs, circles, and splines. In 3D wireframe modeling, the entities such as lines and curves represent the object in Euclidean three-space rather than the projections of edges onto a 2D medium [29]. The wireframe modeling is the least complex method for representing the 3D images as compared to the surface and solid modeling [29]. However, the primary disadvantage with both 2D and 3D wireframe models is that they provide ambiguous representation of the given object. The wireframe models can be used for preview purpose [26]. For example, the rendering of a complex model or an animation sequence with surface or solid modeling can be very time consuming. However, with wireframe modeling, the presentation of complex objects can be done efficiently. Therefore, it can be concluded here that the wireframe models require lesser computer memory space (or storage) and are easier to handle as compared to surface or solid models. Since the wireframe models do not have surface information, they have to be converted into surface models for different operations such as structural analysis and process planning [30]. The advantages and disadvantages of the wireframe model can be discussed as follows:

- *Advantages*:
  - Simple to construct
  - Requirement of less system memory space
  - Lesser processing time
  - Efficient visualization of the objects
  - Easier loading and unloading of the objects
- *Disadvantages*:
  - Complex objects with many edges can be confusing
  - Insufficient information in the model
  - Volume and mass calculations cannot be performed
  - Their visualization depends on the human interpretation

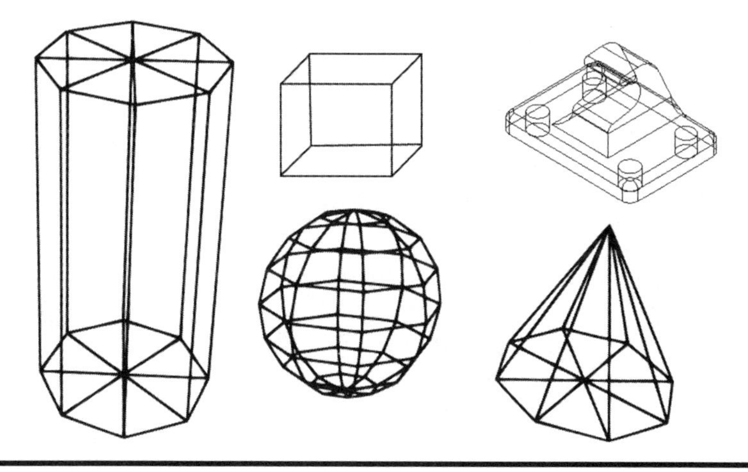

Figure 2.3 2.5 D type model. (Adapted from C. E. Imrak, Three dimensional design and solid modelling, *Computer Aided Technical Drawing—MAK 112E-4,* 2002, course lectures, Istanbul Technical University, Turkey.)

It is worth noticing here that 2.5D types can be defined as the wireframe representations, where the original 2D shape can be translated or rotated into a 3D shape as shown in Figure 2.3.

2.3.2 Surface Modeling

Surface models can be defined as the 3D models with no thickness [34]. In fact, a surface model can be defined as a set of faces [35] as shown in Figure 2.4. To overcome the limitations of the wireframe modeling, surface modeling has been introduced for CAD users where a bounding surface (such as cylindrical, conical, and spherical faces) is generated through the edges and vertices. A surface provides a complete description of the object as compared to the wireframe modeling. These surfaces can be generated using commands such as lofts, sweeps, NURBS curves, etc. [36].

Surface modeling is very useful for creating technical surfaces such as airplane wing and esthetic surfaces such as the car's hood. Moreover, they can be used to define very complex geometries found in forged and molded parts industries,

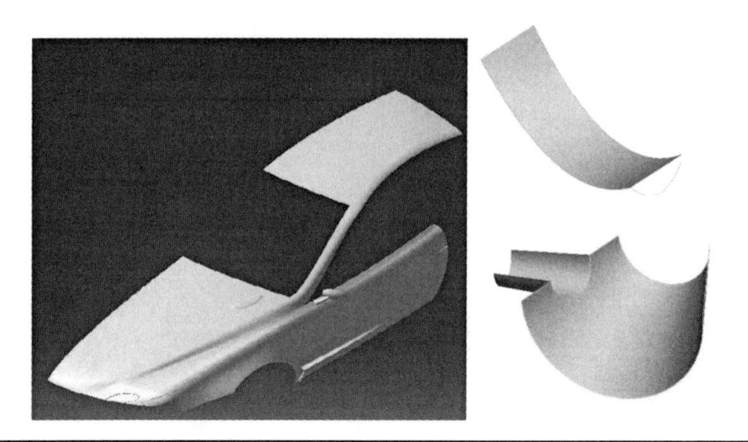

Figure 2.4 Representation of objects using surface modeling. (Adapted from S. Tickoo, *Pro/ENGINEER Wildfire 5.0 for Designers*, CADCIM Technologies, USA, February 12, 2010, 784pp.)

sculptured surfaces, among others. The surface models have widespread applications in industries such as automobile, aerospace, plastic, and medicine. These models can automatically estimate volume property using surface definition but it cannot extract the mass property and other important attributes for manufacturing [23]. Primarily, there are three types of surfaces that can be generated using surface modeling [37]:

1. *Ruled or extruded surfaces*: They are created by skinning two 2D closed curves or extruding one 2D closed curve. It finds its applications in piping design and airplane wing design.
2. *Surfaces of revolution*: The surface of revolution can be generated by rotating the 2D curve about an axis. It used in creating symmetrical objects such as a cylindrical geometry.
3. *Sculptured surfaces*: The sculptured surface is used to generate very complex surfaces such as a ship's hull, an automobile's fender, etc.

There have been several methods, including polygon meshes and parametric cubic patches, to generate the 3D surfaces. Although the algorithm for polygon meshes is simple, they result in the incomplete definition of the curve. The parametric cubic patches determine the coordinates of points on a curved surface using three parametric equations. The bicubic patch is constructed from the corresponding curves such as Ferguson's, Bezier, and the B-spline methods [38].

2.3.2.1 Ferguson's Curve

Ferguson's curve can be defined by two end points (A and B) and two tangents (T_A and T_B) at the endpoints as shown in Figure 2.5.

The Feguson's curve can be defined by the following equations [39]:

$$r(u) = [1 \quad u \quad u^2 \quad u^3] \begin{bmatrix} 1 & 0 & 0 & 0 \\ 0 & 0 & 1 & 0 \\ -3 & 3 & -2 & -1 \\ 2 & -2 & 1 & 1 \end{bmatrix} \begin{bmatrix} A \\ B \\ T_A \\ T_B \end{bmatrix} 0 \le u \le 1$$

2.3.2.2 Bezier's Curve

The Bezier curves allow efficient storage and editing (stretching, rotation, distortion, etc.) of smooth shapes. They can also be used to represent alphabetical letters in various fonts.

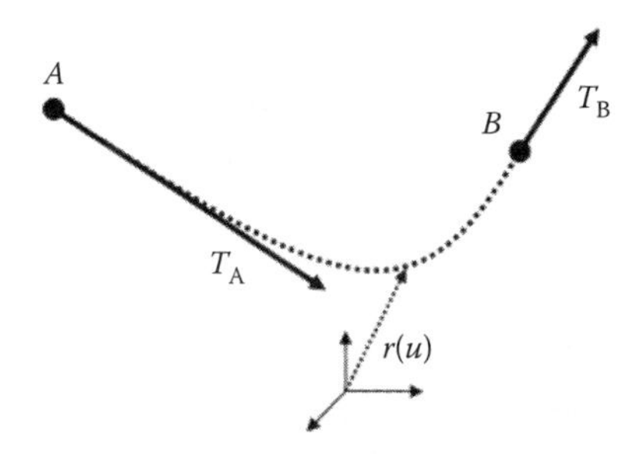

Figure 2.5 Ferguson's curve. (Adapted from A. Sobester and A. J. Keane, Airfoil design via cubic splines - Ferguson's curves revisited. In, *AIAA Infotech@ Aerospace 2007 Conference and Exhibit,* Rohnert Park, USA, 07–10 May. American Institute of Aeronautics and Astronautics. 2007, pp. 1–15.)

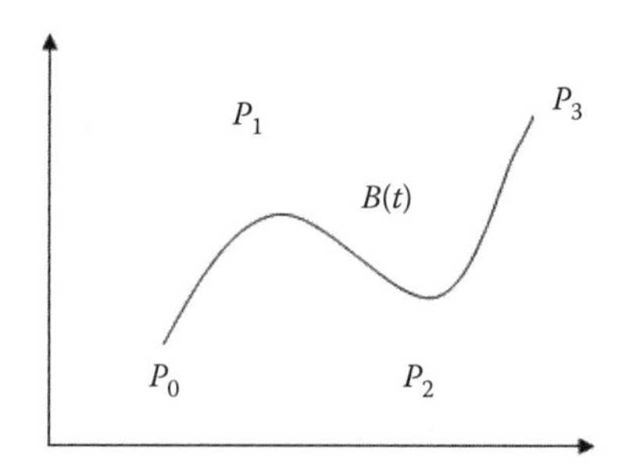

Figure 2.6 Bezier curve *B(t)*. (Adapted from H. P. Wang and R. A. Wysk, *Computer Aided Manufacturing*, 2nd ed. Upper Saddle River, NJ: Prentice-Hall, 1998.)

The Bezier's curve $B(t)$ defined for the four points P_0, P_1, P_2, and P_3 as shown in Figure 2.6 can be defined as follows [38]:

$$B(t) = [P_3 \quad P_2 \quad P_1 \quad P_0] \begin{bmatrix} 1 & 0 & 0 & 0 \\ -3 & 3 & 0 & 0 \\ 3 & -6 & 3 & 0 \\ -1 & 3 & -3 & 1 \end{bmatrix} \begin{bmatrix} t^3 \\ t^2 \\ t \\ 1 \end{bmatrix} 0 \le t \le 1$$

The four points P_0, P_1, P_2, and P_3 are called control points for the Bezier's curve.

Bezier curves are used in computer graphics to produce curves that are reasonably smooth at all scales as compared to polygonal line, which do not scale nicely. Mathematically, they define a special case of cubic Hermite interpolation while polygonal lines use linear interpolation [40]. The advantage of Bezier surface patches is that they are easy-to-sculpt natural surfaces.

2.3.2.2.1 Properties of Bezier Curves

Bezier curves are very useful in design due to the fact that they exhibit the following properties [23,38]:

- $B(0) = P_0$ and $B(I) = P_3$. So, the Bezier curve goes through the points P_0 and P_3. This property ensures that $B(t)$ goes through specified points. For the two Bezier curves to fit together, the value at the end of one curve must match the starting value of the next curve. This suggests that the endpoint values of the Bezier curves can be controlled by selecting the appropriate values for the control points P_0 and P_3.

- $B(t)$ is a cubic polynomial. $B(t)$ is continuous and differentiable at each point, therefore, its graph is connected and smooth at each point. This property also ensures that the graph of $B(t)$ does not distort between control points.
- $B'(0)$ = slope of the line segment from P_0 to P_1:
$B'(1)$ = slope of the line segment from P_2 to P_3. It is impossible to match the ending slope of one curve with the starting slope of the next curve to result in a smooth connection.
- For $0 \le t \le 1$, the graph of $B(t)$ exists in a region whose corners are defined by the control points.
For example, if a rubber band is put around the four control points P_0, P_1, P_2, and P_3 as shown in Figure 2.7, then the graph of $B(t)$ will lie inside the rubber banded region. This property of Bezier curves t guarantees that the graph of $B(t)$ does not get too far from the four control points.

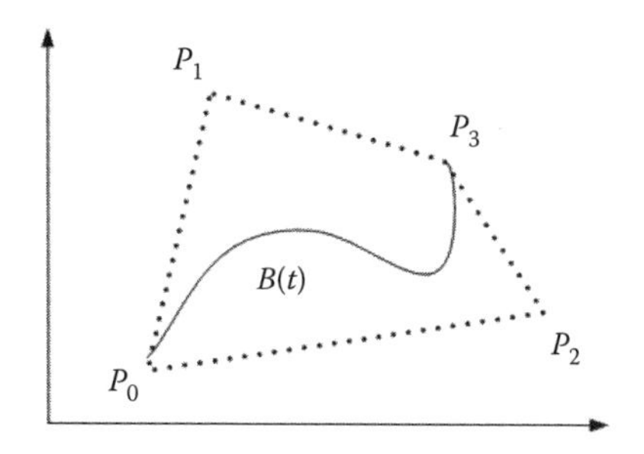

Figure 2.7 $B(t)$ lying inside the region defined by the control points. (Adapted from H. P. Wang and R. A. Wysk, *Computer Aided Manufacturing*, 2nd ed. Upper Saddle River, NJ: Prentice-Hall, 1998.)

2.3.2.3 B-Spline Curve

The nonuniform B-spline curve represents a general case of Bezier's curve. It is usually defined by the Cox–deBoor recursive function [38]:

$$r(t) = \sum_{i=0}^{L} N_i^n(t) p_i \quad t_i \le t \le t_{i+1}$$

$$N_i^n(t) = \frac{t - t_i}{t_{i+n-1} - t_i} N_i^n(t) + \frac{t_{i+n} - t}{t_{i+n} - t_{i+1}} N_{i+1}^{n-1}(t)$$

where

$$N_i^n(t) = \begin{cases} 1, t \in [t_i, t_{i+1}] & \text{and} \quad t_i < t_{i+1} \\ 0, \text{otherwise} \end{cases}$$

and

L = number of control points
n = degree of the curve

The Bezier (Figure 2.8a) and B-spline (Figure 2.8b) methods share several benefits as follows:

- The control points can be adjusted in a predictable way, which make them ideal for use in an interactive CAD environment.
- The local control of the curve shape is possible.

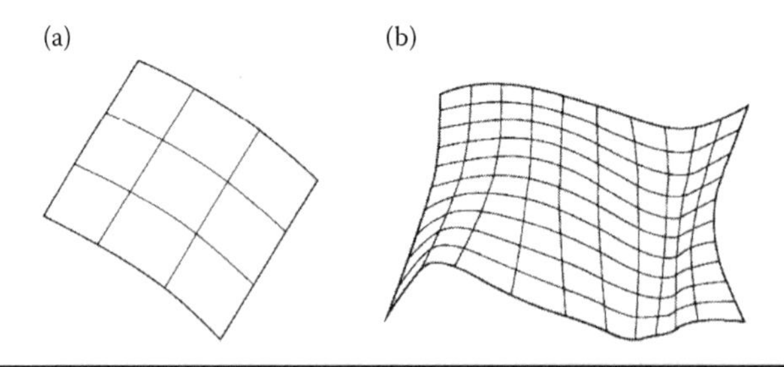

Figure 2.8 (a) Bezier curve and (b) B-spline. (Adapted from C. E. Imrak, Three dimensional design and solid modelling, *Computer Aided Technical Drawing—MAK 112E-4,* 2002, course lectures, Istanbul Technical University, Turkey.)

Therefore, the Bezier method has a disadvantage because if the order of polynomial is increased by adding more control points when more control of shape is needed, the order of polynomial does not change in the B-spline method.

B-spline patches allow local control, which means moving one control point does not affect the whole surface. With B-splines, it is much easier to create surfaces through predefined points or curves. NURBS surfaces use rational B-splines, which include a weighting value at each point on the surface.

2.3.3 Solid Modeling

A solid model as compared to the wireframe and surface models represents the object in a more complete manner. It includes the edges and surfaces and also the volume enclosed by the surfaces as shown in Figure 2.9 [29].

Solid modeling overcomes the limitations of both wireframe and surface modeling by providing a comprehensive solid definition to a 3D object [23]. This technique represents the solid object as a volumetric description, including both the

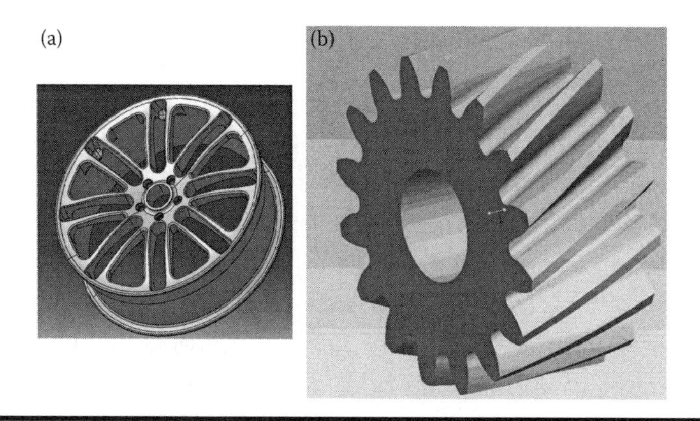

Figure 2.9 Solid modeling. (a) Solid model of a wheel rim obtained using CATIA. (Courtesy of GrabCAD, https://grabcad.com/). (b) Solid model of a gear designed using Pro-E. (Courtesy of Engineer Blogs, http://engineerblogs.org/.)

surface and edge definitions of an object. The mass, volume, and surface properties can be derived using solid models [43]. Moreover, it provides topological information in addition to the geometrical information, which helps to represent the solid unambiguously. The emergence of solid modeling systems allowed designers to develop virtual prototypes or software for visualization and analysis. There are basically two techniques including constructive solid geometry (CSG) and boundary representation (B-rep) for the representation of models in solid modeling [29].

2.3.3.1 History and Overview

In the beginning, the electronic drafting and wireframe models were used to represent the shape of 3D objects. Moreover, the new systems introduced in the 1960s utilized polygonal and surface-based models in a variety of applications in aerospace, marine, and automotive industries. However, the further developments in CAD/CAM systems led to the introduction of solid modeling. The primary reason for the introduction of solid modeling was ambiguity, insufficient information, incomplete or unsatisfactory results, obscure data associated with the wireframe, and surface models. In the late 1970s, most of the issues associated with earlier systems were worked out by the Production Automation Project at the University of Rochester, where the concept of "solid modeling" was established [44]. New mathematical models for representing the solids and mathematical operations were established to manipulate these models. Subsequently, in the 1980s, many solid modeling systems were introduced and used in the commercial CAD/CAM environment in the automobile, aerospace, and manufacturing industries. Furthermore, several advanced CAD/CAM applications such as feature and constraint-based modeling, automatic mesh generation for finite-element analysis, assembly planning, including interference checking, higher dimensional modeling for robotics and collision avoidance, tolerance modeling, and automation of process planning tasks have also surfaced with the development of solid modeling [23].

2.3.3.2 Types of Solid Modeling

The different types solid modeling representation can be described as follows:

- *Cell decomposition*: This technique is used in structural modeling and is the basis of the finite-element modeling [45]. Any solid can be divided into a group of cells whose union or addition results in a representation of the solid itself. The disjoint cells can be of any shape and size.
- *Spatial occupancy enumeration*: It is a special case of cell decomposition where the cells are cubical in shape and exists in a fixed special grid [45]. Here, the object is represented by a list of the cubical disjoint spatial cells that it occupies. It requires large amounts of store for reasonable resolution and therefore has not been favored for practical systems.
- *Primitive instancing*: The object is represented by a set of solid primitives such as cuboid, cylinder, cone, etc. Each primitive is usually defined parametrically and is located in space [46]. A given solid is represented completely by providing the family to which it belongs together with a limited set of parameter values.
- *Sweeping*: The object is represented by moving a curve or a surface along some paths. This method is useful to model constant cross-sectional parts and symmetrical parts [47]. In fact, a solid is defined using volumes swept out by 2D or 3D as they move along a curve.
- *Boundary representation (B-rep)*: The object is enclosed by a set of the bounded faces where each face is represented by its bounding edges and vertices.

- *Constructive solid geometry (CSG)*: The object is represented by combining several primitive shapes through Boolean operations.

2.4 Boundary Representation (B-Rep)

Boundary representation (B-rep) methods are used to define the bounding surfaces of a solid object [29] as shown in Figure 2.10. A B-rep solid is represented as a volume enclosed by a set of faces along with the topological information, which defines the relationships between the faces [48]. The advantage of B-rep is that the nongeometric data, such as tolerances, can be assigned to the geometric entities [49]. However, it is very difficult to determine the specific feature in a pure B-rep model.

B-rep defines the geometry of the object using the boundaries, that is, the vertices, edges, and surfaces, which represent entities of two dimensions, one dimension, and zero dimension, respectively [50]. In B-rep, the orientation of each of the surface has to be defined as the interior or the exterior in order to represent the solid object using surfaces. Moreover, in a B-rep, a face should fulfil the following conditions:

- The finite number of faces should define the boundary of the solid.

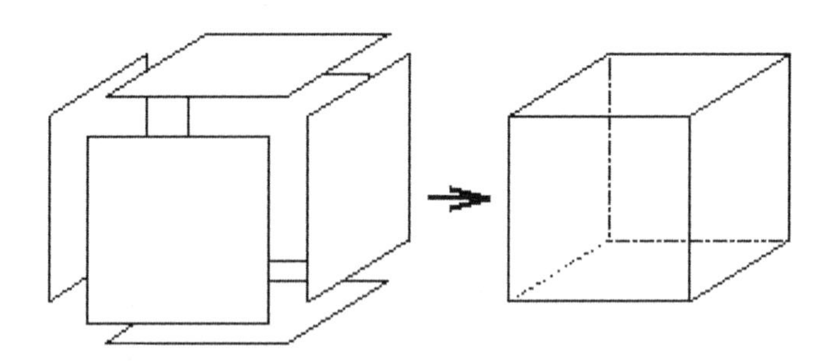

Figure 2.10 B-rep model created using six faces. (Adapted from I. Zeid and R. Sivasubramanian, *CAD/CAM-Theory and Practice*, 2nd Edition, New York: Tata McGraw-Hill Education, 1991, 813pp.)

- The face of a solid must be a subset of the solid's boundary.
- The union of all the faces should define the boundary.
- The face itself should be limited region or a subset of the more extensive surface.
- A face must have a finite area and is dimensionally homogeneous.

The B-rep models are not limited to a planar surface, however, different types of surface geometries such as curved surfaces can be represented. Moreover, the B-rep models can represent a large group of objects and are ideal choice for modeling the complex objects. The B-rep solids modeling require more memory space as compared to CSG because it takes a large number of points to define each surface. Moreover, the application of Boolean operations in B-rep models is limited by the complexity of the object.

2.4.1 Euler's Formula

The validity of the B-rep model can be verified by using Euler's formula [51]. In a topological valid B-rep, each edge is always exactly adjacent to the two faces and is terminated by two vertices.

Euler's formula states that for any topological valid solid, the number of vertices in addition to the number of faces is equal to the number of edges plus two.

$$\boxed{V + F = E + 2,}$$

where

V = number of vertices
F = number of faces
E = number of edges

The validity of the resultant model (Figure 2.11) obtained by combining the two B-rep models can be checked using Euler's formula.

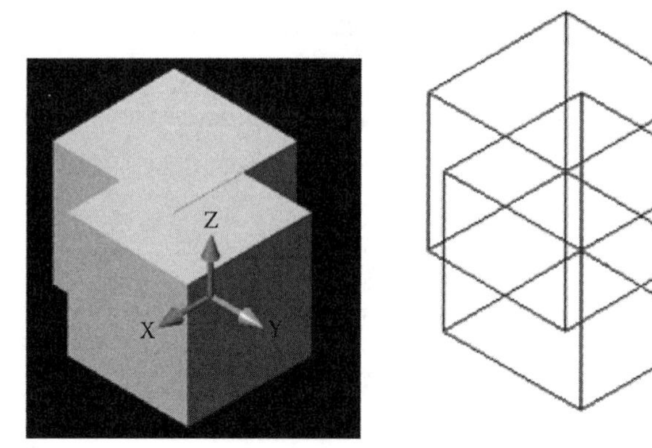

Figure 2.11 Combined B-rep models.

From the model, it can be concluded that

Total number of vertices, $V = 20$
Total number of faces, $F = 12$
Total number of edges, $E = 30$

Applying Euler's formula:

$$V + F = 20 + 12 = 32$$
$$E + 2 = 30 + 2 = 32$$

Therefore, $V + F = E + 2$

This suggests that the resultant model has the correct number of topological elements.

2.5 Constructive Solid Geometry (CSG)

The CSG is a modeling method that defines the complex solids as compositions of simple solid primitives [44]. In fact, CSG can be defined as a method of building up complex solids (shown in Figure 2.12b) through the application of various Boolean operations on a set of primitive solids (Figure 2.12a). The common CSG operations are union, intersection, and difference.

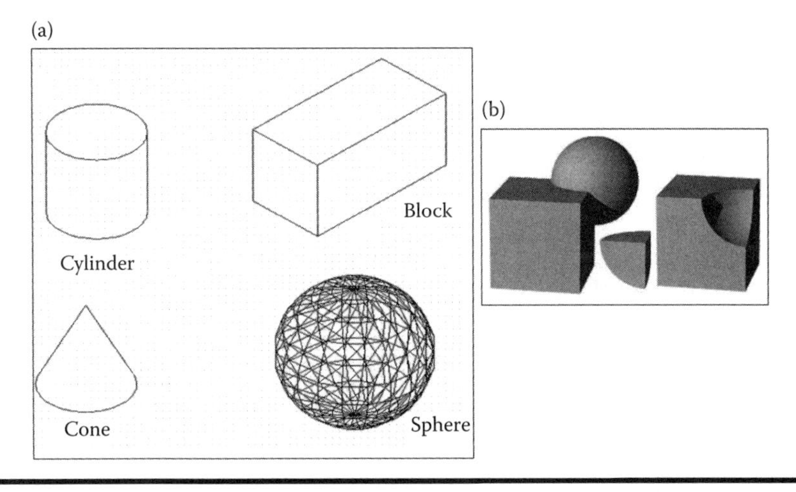

Figure 2.12 (a) Primitive solids. (Adapted from D. A. Friis, Introduction to Computer Aided Design Drafting, Memorial University of Newfoundland, http://www.engr.mun.ca/~dfriis/cadkey/program/textch1.html). (b) Constructive solid geometry. (Courtesy of Fuzzy Photon by Siddhartha Chaudhuri, 2002, http://fuzzyphoton.tripod.com/.)

The CSG is efficient because it requires lesser computer memory space to completely define an object. For example, a cylinder can be represented by its primitive type, height, diameter, location space, and material density. However, it is very difficult to define complex shapes such as automobile hood, aircraft wing, ship hulls, etc. using CSG. In the CSG model representation, the part design is represented by an ordered binary tree. The ordered binary tree consists of nodes, where these nodes are either terminal or nonterminal [54].

2.6 Advantages and Disadvantages of CSG and B-Rep [23,55]

- *Advantages of CSG*:
  - Concise and compact representation
  - More user-friendly and easier to implement
  - Established algorithms for conversion of CSG into B-rep

- The data structure is relatively simple, robust, and easy to handle
- Relatively easy to modify and can be easily converted into a B-rep
- Easy analysis (rendering, mass properties, etc.).

- *Disadvantages of CSG*:
  - The database consists of information in an unevaluated form.
  - The validity of a feature of an object cannot be assessed without evaluating the entire tree.
  - Representation is not unique. The tree is not unique for the same part design.
  - No adjacency information (difficult to distinguish feature).
  - No direct access to vertices (difficult for designer to select specific part of the object).
  - Hard to extract important features for manufacturing (holes, slots, etc.).

- *Advantages of B-rep*:
  - Adjacency information, therefore, easy to distinguish feature.
  - The information is complete, especially for adjacent topology relations.
  - B-rep has the most refined geometric information.
  - B-rep is more flexible and has a much wider operation set. It has extrusion, chamfering, blending, drafting, shelling, tweaking, and other operations.
  - Unique representation.
  - Verification of model correctness using Euler's formula.
  - Tweaking of vertices and edges is possible.
  - Easy to extract important features for manufacturing.

- *Disadvantages of B-rep*:
  - Large database
  - B-rep requires feature extraction procedures to extract features from its face-edge-vertex database

- The B-rep model does not provide any explicit information, spatial constraints between features
- Complex data structure of B-rep
- Slow computation of B-rep
- Difficult to maintain robust models for curved surfaces
- Difference between geometry and topology

2.7 Feature Recognition

Feature recognition can be defined as the identification and grouping of feature entities from a geometric model. In fact, it extracts features and their parameters from the solid models. Generally, the identified entities (i.e., the recognized features) are extracted from the model and engineering information such as tolerances and nongeometric attributes are then attached to the feature entities [56]. The flow diagram of the feature recognition can be seen in Figure 2.13. There are a number of factors that make feature recognition an indispensable part of the CA/CAM systems [57].

- Efficient utilization of CAD data in downstream applications
- Nonexistence of industry standard for feature definition and storage
- Solid models without feature information cannot be easily edited without using feature recognition systems
- Since there are various types of features such as the design feature or manufacturing depending on the application requirement, feature recognition systems are required to identify the features for different applications.
- In CAE such finite-element analysis (FEA), there is a need to recognize and inhibit the unwanted features and speed the analysis process.

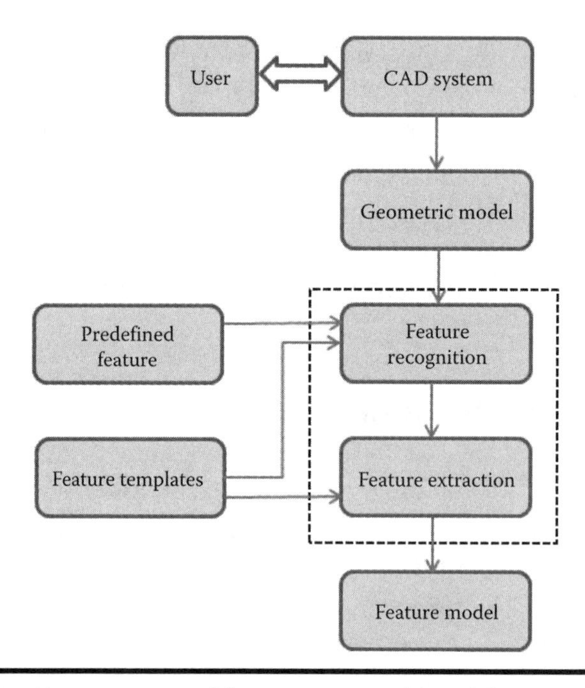

Figure 2.13 Feature recognition system. (Adapted from E. A. Nasr and A. K. Kamrani, *Computer-Based Design and Manufacturing: An Information-Based Approach*. Berlin: Springer, 2007.)

2.8 Feature-Based Design

Design by features, or the feature-based design (FBD), can be defined as a design methodology where a library of 2D or 3D features (as design primitives) is utilized to model a given product. This is due to the fact that the use of features provides a more natural interface between design and solid model [58]. For example, a designer can work directly with high-level entities such as a pocket or slot rather than utilizing low-level entities in which vertices and edges form a pocket. Moreover, the features provide the capability of associating additional information useful for process planning [59]. The overview of the feature-based design methodology can be seen in Figure 2.14.

The integration of the design by feature and feature recognition is feasible for the integration of design and

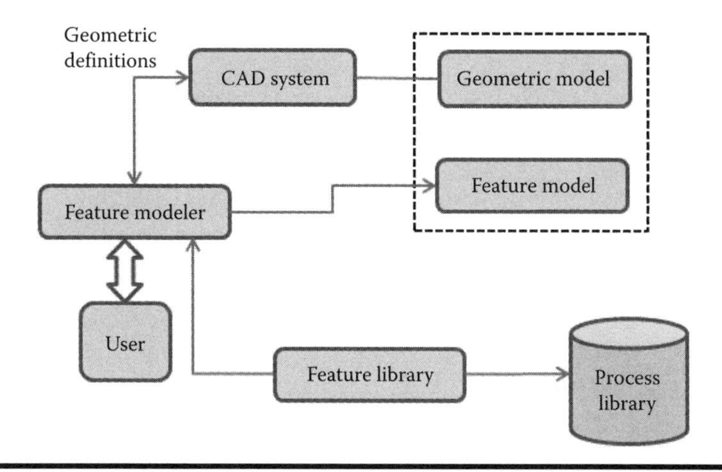

Figure 2.14 Feature-based design methodology. (Adapted from E. A. Nasr and A. K. Kamrani, *Computer-Based Design and Manufacturing: An Information-Based Approach.* Berlin: Springer, 2007.)

manufacturing processes. It is easy to derive a geometric feature from the higher level entities of a part model, in which each feature is associated with a component of a solid model. On the other hand, to transform the geometric model into a feature representation is more difficult [23].

2.9 Feature Interactions

Feature interactions can be defined as the intersections of feature boundaries with those of other features such that either the shape or the semantics of a feature are altered from the standard or generic definition [10]. It is important here to explain the difference between the interacting and the intersecting features. The interacting features are defined in the context of design by features in which design features are used to build the model [60]. The addition (or subtraction) of a feature to the part can result in the generation of several new features due to interaction between the features. On the contrary, the intersecting features are present on conventional

(a)

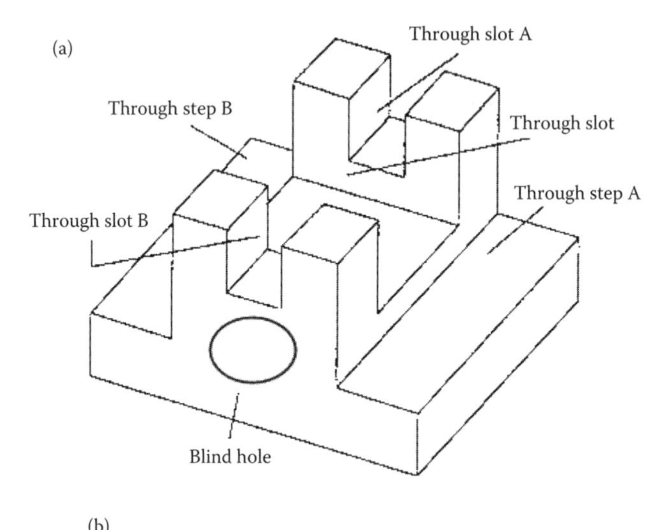

(b)

Feature	Type
Simple	Blind hole
Interacting	Through step A interacting through slot
	Through step B interacting through slot
	Through slot A interacting through slot
	Through slot B interacting through slot

Figure 2.15 (a) Component with interacting features and (b) interacting features. (Adapted from E. A. Nasr and A. K. Kamrani, *Computer-Based Design and Manufacturing: An Information-Based Approach.* Berlin: Springer, 2007.)

representations and not generated by feature operations [58]. There exists no feature recognition system that can recognize all types of 3D solid primitive features and their various interacting combinations [61,62]. There are six features on the raw material of the workpiece as shown in Figure 2.15a and the interaction between the features can be seen in Figure 2.15b.

2.10 Summary

The concept of geometric modeling, including wireframe modeling, surface modeling, boundary representation (B-rep), constructive solid geometry (CGS), and definition of interacting features has been introduced in this chapter. It also provides a good description of feature definitions, process planning, feature recognition, and assembly planning.

QUESTIONS

1. Define “feature.” Classify the different features.
2. What do you mean by implicit and explicit features?
3. What are the characteristics of a feature?
4. Write down the difference between the form features and primitive features.
5. What is the difference between the design feature and machining feature?
6. How is feature-based modeling significant to manufacturing?
7. Define process planning. What are the primary approaches used in process planning?
8. What is the difference between variant and generative process planning?
9. Write down the advantages and disadvantages of variant and generative process planning.
10. Write down the different steps involved in variant process planning.
11. What is the difference between preparation and production stages of the variant process planning?
12. Explain the major components of the generative process planning.
13. Describe the different phases of assembly planning.
14. How is inspection planning important in the manufacturing process?
15. Describe geometric modeling. What are the different methods used in geometric modeling?
16. Explain wireframe modeling. Write down its various advantages and disadvantages.
17. Define 2.5 types of wireframe model.
18. Define surface modeling. What are its advantages and disadvantages?
19. Describe the types of surfaces that can be generated using surface modeling techniques.

20. Explain the following:
    a. Ferguson curve
    b. Bezier's curve
    c. B-spline curve
21. Write down the various properties of Bezier's curve.
22. Explain solid modeling. What are the different types of solid modeling techniques?
23. Write down the difference between boundary representation (B-rep) and constructive solid geometry (CSG).
24. Describe Euler's formula. If a resultant model has a total of 20 vertices, 12 faces, and 30 edges, verify whether the resultant model has the correct number of topological elements.
25. Write down the advantages and disadvantages of B-rep and CSG models.
26. What are the criteria that a face must fulfil in a B-rep model?
27. What do you mean by feature recognition? Why is it an integral part of the CAD/CAM system?
28. Explain the following:
    a. Feature-based design
    b. Feature interactions

References

1. M. C. Kayacan and Ş. A. Çelik, Prizmatik Parçaların Bilgisayara Tanıtılmas, *MAMKON'97, İTÜ Makina Fakültesi 1. Makina Mühendisliği Kongresi*, İstanbul, 1997;122–133.
2. H. Başak and M. Gülesin, A feature based parametric design program and expert system for design, *Mathematical and Computational Applications*, 2004;9(3):359–370.
3. M. Shpitalni and A. Fischer, CSG representation as a basis for extraction of machining features, *Annals of the CIRP*, 1991;40(1):157–160.
4. S. Subrahmanyam and M. Wozny, An overview of automatic feature recognition techniques for computer-aided process planning, *Computers in Industry*, 1995;26:1–21.

5. R.-F. Wang and J. Turner, Recent research in feature based design, Technical Report No. 89020, Rensselaer Design Research Center, Rensselaer Polytechnic Institute, Troy, NY, May 1989.
6. F. Giacometti and T. C. Chang, A model for parts, Assembly and tolerance, *IFIP W.G.5.2 Workshop on Design for Manufacturing*, 1990.
7. R. Sodhi and J. U. Turner, Representing tolerance and assembly information in a feature based design environment, *Advances in Design Automation*, DE 1991;32-1, ASME:101-106.
8. T. Kiriyama, T. Tomiyama, and H. Yoshikawa, The use of qualitative physics for integrated design object modelling, *ASME Conference on Design Theory and Methodology*, December 1991;31:53–60.
9. J. J. Shah, Conceptual development of form features and features modelers, *Research in Engineering Design*, 1991;2:93–108.
10. J. J. Shah and M. Mantyla, *Parametric and Feature-Based CAD/CAM*, New York: Wiley, 1995.
11. H.-C. Zhang and L. Alting, *Computerized Manufacturing Process Planning Systems*, London: Chapman & Hall, 1994.
12. S. M. Amaitik, Development of a STEP feature-based intelligent process planning system for prismatic parts, PhD Thesis, Middle East Technical University, Mechanical Engineering, April 2005.
13. J. J. Shah and M. T. Rogers, Expert form feature modeling shell, *Computer-Aided Design*, 1988;20(9):515–524.
14. J. Shah, *Parametric and Feature-Based CAD/CAM—Concepts, Techniques, and Application*. New York: Wiley, 1995.
15. S. Yao, Computer-aided manufacturing planning (CAMP) of mass customization for non-rotational part production, PhD Dissertation, Worcester Polytechnic Institute, December 2003.
16. N. Wang and T. M. Ozsoy, A scheme to represent features, dimensions, and tolerances in geometric modeling, *Journal of Manufacturing Systems*, 1991; 10(3):233–240.
17. J. Y. Lee and K. Kim, Generating alternative interpretations of machining features, *International Journal of Advanced Manufacturing Technology*, 1999;15:38–48.
18. N. Ahmad, A. F. M. Haque, and A. A. Hasin, Current trend in computer aided process planning, *Proceedings of the 7th Annual Paper Meeting and 2nd International Conference of the Institution on Engineering*, Dhaka, Bangladesh, 2001;1(10):81–92.

19. I. Ham and S. C.-Y. Lu, Computer-aided process planning: The present and the future, *CIRP Annals—Manufacturing Technology* 1988;37(2):591–601.
20. H. B. Marri, A. Gunasekaran, and R. J. Grieve, Computer-aided process planning: A state of art, *The International Journal of Advanced Manufacturing Technology*, 1998;14(4):261–268.
21. M.P. Groover, *Automation, Production Systems, and Computer-Integrated Manufacturing*. Englewood Cliffs, NJ: Prentice-Hall, 2001.
22. B. Beno, *Manufacturing Design, Production, Automation and Integration*, New York: Marcel Dekker, 2003.
23. E. A. Nasr and A. K. Kamrani, *Computer-Based Design and Manufacturing: An Information-Based Approach*. Berlin: Springer, 2007.
24. T. N. Goh and M. Xie, Improving on the six sigma paradigm, *The TQM Magazine*, 2004;16(4):235–240.
25. O. Bartholomew, B. O. Nnaji, and H. C. Liu, Feature reasoning for automatic robotic assembly and machining in polyhedral representation, *International Journal of Production Research*, 1990;2(3):517–540.
26. S. C. Agarwal and W. N. Waggenspack Jr., Decomposition method for extracting face topologies from wire frame models, *Computer-Aided Design*, 1992;24(3):123–140.
27. M. P. Groover and E. W. Zimmers, *CAD/CAM: Computer Aided Design and Manufacturing*, Englewood Cliffs, NJ: Prentice-Hall, 1984, 489pp.
28. M. Mortenson, M. *Geometric Modeling*, 1st ed., New York: Wiley, 1985.
29. K. Holly, Ault, 3-D Geometric modeling for the 21st Century, *Engineering Design Graphics Journal*, 1999;63(2):33–42.
30. P. Radhakrishnan, S. Subramanyan, and V. Raju, CAD/CAM/CIM, New Age International, CAD/CAM systems, 2008, 688pp.
31. C. E. Imrak, Three dimensional design and solid modelling, *Computer Aided Technical Drawing—MAK 112E-4*, 2002, course lectures, Istanbul Technical University, Turkey.
32. J. Rooney and P. Steadman, *Principles of Computer-Aided Design*, London: UCL Press, 1997.
33. O. Ostrowsky, *Engineering Drawing with CAD Applications*, Routledge, UK, July 3, 1989.
34. S. Tickoo, *Pro/ENGINEER Wildfire 5.0 for Designers*, CADCIM Technologies, USA, February 12, 2010, 784pp.

35. A. Tura and Z. Dong, Solid Modelling, Computer Aided Design, University of Victoria, http://www.engr.uvic.ca/~mech410/old/2_Lecture_Notes/5_Geometric_Modeling.pdf (accessed on June 22, 2015).
36. The Hongkong Polytechnic University, Solid Modelling (Chapter 6), Computer Modelling, Industrial Centre, 28pp, http://fireuser.com/articles/solids_vs_surface_modeling_what_and_why_you_need_to_know/ (accessed on June 22, 2015).
37. S. H. Chasen, Principles of geometric modeling, *CIM Technology*, 15–18, 1986.
38. H. P. Wang and R. A. Wysk, *Computer Aided Manufacturing*, 2nd ed. Upper Saddle River, NJ: Prentice-Hall, 1998.
39. A. Sobester and A. J. Keane, Airfoil design via cubic splines—Ferguson's curves revisited, *AIAA Infotech@ Aerospace 2007 Conference and Exhibit*, Rohnert Park, USA, 07–10 May. American Institute of Aeronautics and Astronautics. 2007, pp. 1–15.
40. B. Casselman, Bezier Curves, Department of Mathematics, University of British Columbia, Vancouver, Canada, https://www.math.ubc.ca/~cass/gfx/bezier.html (accessed on June 23, 2015).
41. GrabCAD, https://grabcad.com/library/sample-car-rim (accessed on June 23, 2015).
42. Engineer Blsogs, http://engineerblogs.org/2011/05/proe-my-love-hate-relationship/ (accessed on June 23, 2015).
43. C. Hoffmann and J. Rossignac, A road map to solid modeling, *IEEE Transactions on Visualization and Computer Graphics*, 1996;2(1):45–54.
44. A. Reqicha and H. Voelcker, Solid modeling: A historical summary and contemporary assessment, *IEEE Computer Graphics and Applications*, 1982;2(2):9–24.
45. C. A. Mota Soares, *Computer Aided Optimal Design: Structural and Mechanical Systems*. Berlin: Springer Science & Business Media, 2012, 1029pp.
46. A. A. G. Requicha, Representations of rigid solids: Theory, methods, and systems, *Computing Survey*, 1980;12(4):437–464.
47. D. Taylor, *Computer Aided Design*, Reading, MA: Addison-Wesley, 1992.
48. T. Várady, R. R. Martin, and J. Cox, Reverse engineering of geometric models—An introduction, *Computer-Aided Design* 1997;29(4):255–268.
49. L. C. Sheu and J. T. Lin, Representation scheme for defining and operating from features, *Computer Aided Design*, 1993;25(6):33–347.

50. T.-C. Chang, *Expert Process Planning for Manufacturing*, Reading, MA: Addison Wesley, 1990.
51. I. Zeid, *CAD/CAM Theory and Practice*, New York: McGraw-Hill, 1991.
52. D. A. Friis, Introduction to Computer Aided Design Drafting, Memorial University of Newfoundland, http://www.engr.mun.ca/~dfriis/cadkey/program/textch1.html. (accessed on June 23, 2015).
53. Fuzzy Photon by Siddhartha Chaudhuri, 2002, http://fuzzyphoton.tripod.com/ (accessed on June 23, 2015).
54. D. L. Waco and Y. C. Kim, Geometric reasoning for machining features using convex decomposition, *Computer Aided Design*, 1994;26(6):477–489.
55. S. N. Shome, J. Basu, G. P. Sinha, *Proceedings of the National Conference on Advanced Manufacturing & Robotics*, January 10–11, 2004 (Computer integrated manufacturing systems, 570 pp).
56. P. C. Sreevalsan and J. J. Shah, Unification of form feature definition methods, Pres. IFIP WG5.2 Wkshp. Intelligent CAD Systems Columbus, OH, September, 1991.
57. W. C. Regli, III, Geometric Algorithms for Recognition of Features from Solid Models. PhD Thesis, The University of Maryland, 1995, http://feature.geometricglobal.com/ (accessed on June 24, 2015).
58. F. L. Wen and M. Ronak, Feature-based design in an integrated CAD/CAM system for design for manufacturability of machining prismatic parts, *Concurrent Product Design and Environmentally Conscious Manufacturing*, 1997;5(1):95–112.
59. M. Shpitalni, CSG representation as a basis for extraction of machining features, CIRP Annuals, *Manufacturing Technology*, 1990;40(1):157–160.
60. Y. J. Tseng and S. B. Joshi, Recognizing of interacting rotational and prismatic machining features from 3D mill-turn parts, *International Journal of Production Research*, 1998;36(11):3147–3165.
61. V. Allada and S. Anand, Machine understanding of manufacturing features, *International Journal of Production Research*, 1996;34(7):1791–1820.
62. K. Huikange, M. Nandakumar, and J. Shah, CAD/CAM integration using machining features, *International Journal of Computer Integrated Manufacturing*, 2002;15(4):296–318.
63. I. Zeid and R. Sivasubramanian, *CAD/CAM-Theory and Practice*, 2nd Edition, New York: Tata McGraw-Hill Education, 1991, 813pp.

Chapter 3

Automated Feature Recognition

Feature recognition can be defined as a process utilized to identify the manufacturing features of the part using its solid model [1]. It plays a significant role in a number of engineering applications and serves as an important support tool for the integrated manufacturing [2]. It aids to accomplish the interface between the CAD and computer aided process planning (CAPP) systems. It is one of the first and the most important steps to carry out the integration of the CAD, CAM, and computer aided inspection (CAI) systems. The main purpose of the feature recognition is to provide a direct link between the CAD and CAM systems so that the integrated productivity of CAD/CAM systems can be improved [3]. The feature recognition can be described as a technique in which a set of rules is applied to extract high-level semantic information from the part model. It consists of three main steps [4]:

1. Feature definition: In this step, the set of rules is defined for the given feature.
2. Feature classification: The important features are classified at this stage.

3. Feature extraction: The features are extracted using the solid model and stored for further study.

The application or the manufacturing features are automatically recognized from the geometric model of the object using the geometric reasoning techniques that actually investigate the data structure of the solid modeler [5]. Since, in the feature recognition, the features are extracted using the geometric model of the part, the designer can define any object shape using the geometric primitives [6]. The feature recognition can be broadly classified as human-assisted feature recognition and the automatic feature recognition [4]. The human-assisted feature recognition systems involve a higher degree of human intervention at all the stages of the recognition process. On the contrary, the recognition and extraction stages are fully automated in the automatic feature recognition systems. Moreover, the whole concept of automated feature recognition is based on a number algorithms to recognize the required form feature [7]. In fact, the automatic feature recognition greatly impacts the level of CAD and CAM integration in the downstream applications such as process planning [2]. The automatic feature-recognition algorithms can be categorized as the machining-region, rule-based, graph-based, CSG-based, and application-based algorithms [4]. Moreover, the difference between the feature recognition and the design by feature approach can be explained as follows [8]. In feature recognition, the solid model of the part is developed and then the manufacturing features are identified (or extracted) from the generated solid model. However, in the design by feature approach, a database of manufacturing features is maintained for the feature extraction. For example, the designer can utilize the features such as holes and slots, to synthesize the part design. In fact, in feature-based design, the components are first designed through a simple extrusion or revolution (such as a cylinder or a block) and then modified by incorporating features such as holes and slots [9].

3.1 Feature Representation

There have been a number of techniques such as constructive solid geometry (CSG), and boundary representation (B-rep), which are used by the solid modelers to represent the part [4]. The B-rep is the most commonly used technique by the various solid modelers, whereas the CSG part representation has found relatively less application [7]. The CSG stores the part information as a tree of primitive volumes, including their respective Boolean operators utilized to construct the solid geometry. However, the B-rep contains the boundary elements of the resultant object. The B-rep database stores low-level primitives such as faces, edges, and vertices and their topological relationships [10]. The robustness of a B-rep justifies its usefulness in industrial applications and, therefore, this type of solid modeling is popular in current industries [10]. Moreover, in B-rep-based systems, the features comprises of the vertices, edges, and faces of the part, while, in CSG-based systems, the geometric primitives, such as sphere, cylinder, and so on are used [7].

3.1.1 Feature Representation by B-Rep

In the B-rep scheme, the solid is modeled using its boundary faces, which are bounded regions on planes or surfaces [3,11]. The boundary face is indirectly defined by its bounding loops, edges, and vertices without any concise mathematical formula for its explicit representation. The B-rep method has the following attributes [5]:

- The object is represented by its boundary.
- The object's boundary is described as the union of its bounding faces and the adjacency relationships between them.
- The faces are expressed in terms of their bounding edges, and the data indicates which portions of the faces constitute the bounding edges.

- B-rep contains explicitly the data of low-level geometric entities, such as faces, edges and vertices.
- It maintains strong adjacency relationships between the various types of geometric entities with pointers.

Moreover, the B-rep of a solid model contains information about the faces, edges, and vertices of the surface model, including topological information, which defines the relationship among the faces, edges, and vertices [12].

3.1.2 Feature Representation by CSG

The CSG-based techniques are based on the CSG representations of the solid models. The CSG rep of a solid model is specified using a set of Boolean operations and a set of 3D primitive solids (e.g., blocks, cylinders, and spheres) [7,12,3]. The solids represented in CSG are implicitly defined, that is, their shape is not known unless the associated Boolean operations are evaluated [3]. A CSG tree is simpler and more concise than a B-rep representation [4]. It also stores the necessary Boolean operations along with the primitives in order for the automatic generation of the process plan. However, the main problem associated with the CSG trees is that they are nonunique [4].

3.1.3 Feature Representation by B-Rep and CSG (Hybrid Method)

The hybrid CSG/B-rep data structure is a useful feature representation method because it can represent both the high-level features and low-level basic geometric entities [13]. It can support both the conventional tolerances as well as the geometric tolerances [13].

Wang and Ozsoy [14] used hybrid scheme successfully to represent the features, dimensions, and tolerances in the geometric modeling. This hybrid representation scheme combining

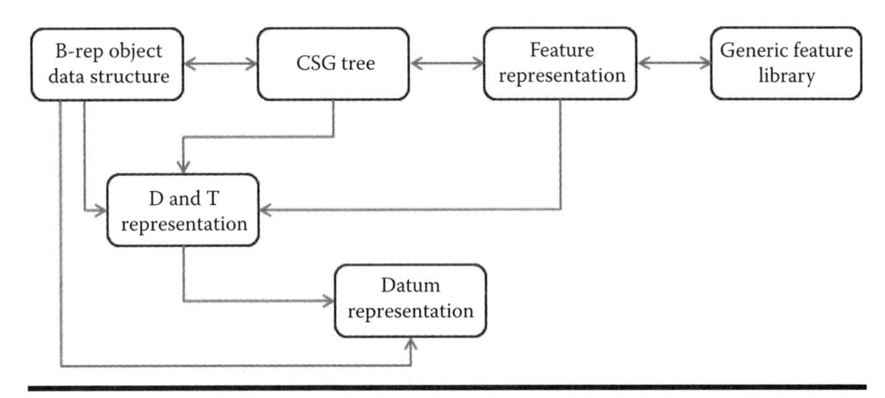

Figure 3.1 Hybrid representation scheme. (Adapted from N. Wang and T. M. Ozsoy, *Journal of Manufacturing System*, 1991;10(3):233–240.)

the CSG tree with a face-edge based B-rep data structure can be seen in Figure 3.1. In this scheme, the primitive features are explicitly defined in the B-rep data structure using the faces, edges, and vertices. These primitive features can be obtained whenever they are required, such as for specifying the dimensions or tolerances. However, the form features are first stored in a generic feature library and when the object is created, the form features are first referred and connected to the CSG tree. Then, it is evaluated and stored in the B-rep structure. The implicit representation of the form feature is attached to the CSG tree and possesses a set of parameters and a set of feature faces to outline the local geometrical and topological relations in the form feature. The explicit representation, which is stored in the B-rep data structure during the evaluation, includes faces, edges, and vertices with each face related to the corresponding feature face. The advantage of this hybrid representation is that it helps the user to modify form features by changing their parameters. The hybrid feature representation scheme provides the following benefits [14]:

- It provides a new design tool, which aids in the creation of a machine part in terms of features and dimensions.
- Any change to the stored object can be achieved by changing the corresponding dimensions.

- The evaluation of the CSG tree involves less computation, resulting in the better interactivity.
- It is useful for the applications such as engineering applications such as tolerance analysis, dimensioning checking, and offline coordinate measuring machine programming, etc.

3.2 Feature Recognition Techniques

3.2.1 The Syntactic Pattern Recognition Approach

In the syntactic pattern recognition method, the model of the part is created using the semantic primitives, which are written in some description language [7]. A set of grammar, which consists of some rules, defines a particular pattern. The working of this approach can be described as follows. The parser for the input sentence analysis is utilized to apply a grammar to the entities defining the part. If the syntax conforms to the grammar, then the description can be categorized into a corresponding class of forms (pattern). There are three components of pattern recognition [7] as shown in Figure 3.2:

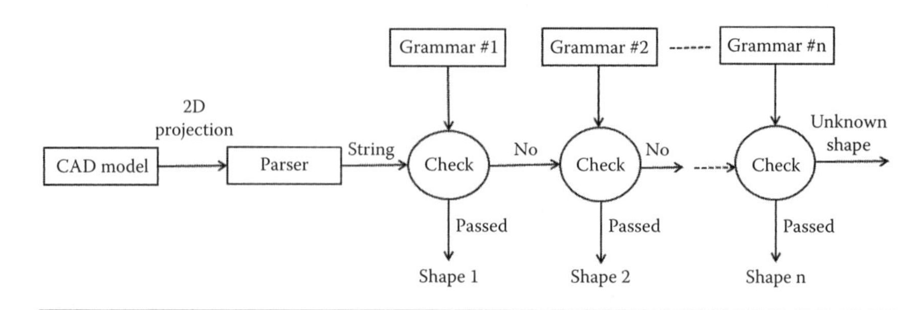

Figure 3.2 The schematics of the syntactic pattern recognition approach. (Adapted from B. Babic, N. Nesic, and Z. Miljkovi, *Computers in Industry,* 2008;59:321–337.)

1. Input string, which represents a semantically unknown grammar.
2. Form semantics are recognized after its classification in a group of predefined forms (pattern). This classification is made through form syntax lookup.
3. Pattern syntax is also defined using grammar.

This method requires the definition of the form primitives and the automated translation of design model, which is suitable for syntax analysis (string) [7].

It is a simple approach, which identifies simple features. Initially, a pattern picture with predefined grammar is considered as shown in Figure 3.3a [2]. Then, a syntactic pattern string is formed for faces by examining the directions of the oriented edges exist in the edge loop of the face. The directions of the oriented edges in the STEP file are represented in a clockwise direction with respect to the normal vector on the face as shown in Figure 3.3b [2].

The main disadvantage of the syntactic pattern recognition is its limited applications. Its applications are limited to the 2D prismatic parts, rotational parts by turning features and axis symmetric volumes [7,9]. Moreover, its application for the nonaxis symmetric 3D part or rotational parts with nonturning

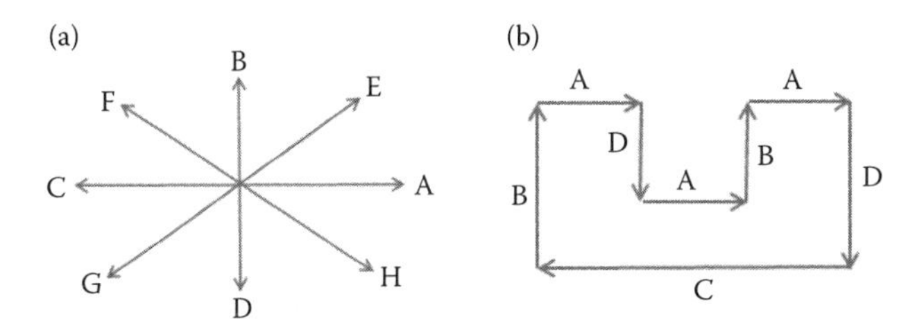

Figure 3.3 (a) Predefined pattern grammar and (b) face and edge loop direction. (Adapted from B. Venu, V. Komma, and D. Srivastava, A New Hybrid Approach to Recognize Machinable Features of Prismatic Parts from STEP AP 203 File, *5th International & 26th All India Manufacturing Technology, Design and Research Conference (AIMTDR 2014),* December 12–14, 2014, IIT Guwahati, Assam, India.)

features has not been established so far. The complexity of this technique increases when it is applied to the prismatic parts, which lacks rotational property. The syntactic pattern recognition handles only the objects that can be described by a string grammar for its 2D cross section [15,16].

3.2.2 The Logic-Based Approach

The logic rules can be used with the B-rep modeling and the CSG modeling approaches for feature recognition [17]. In this approach, the features are recognized, one by one in sequential manner until the complete set of features is obtained for the given solid model. For example, the following set of heuristic rules can be used to describe a slot feature [3].

- A slot is composed of three faces $\{F_1, F_2, F_3\}$
- Face F_1 is adjacent to face F_2.
- Face F_2 is adjacent to face F_3.
- Face F_1 and F_3 are parallel.
- Face F_1 forms a 90° angle with F_2.
- Face F_3 forms a 90° angle with F_2.

Similarly, to recognize the pocket feature, the following rules can be applied [17]:

IF	face is adjacent to face and
	face is adjacent to face and
	face is adjacent to face and
	face is adjacent to faces and
	angle between and is <180 (concave),
	and
	angle between and is <180 (concave)
THEN	faces and form a pocket feature

The following rule can be implemented for hole recognition in CSG representation [17]:

IF	the Boolean operation is subtraction,
	and the dimensions of the subtracted solid
	primitive are less than the solid model
	and the subtracted solid primitive is a cylinder
THEN	the feature is a hole

This approach is based on the algorithms that identify a feature on the basis of certain prespecified rules, which are characteristic to the feature. It can be used to recognize features such as holes, fillets, and bosses, which are very limited in nature [18]. Although this type of approach can deal with general 3D shape patterns, still, writing a rule for every specific pattern is difficult and time-consuming [19–21]. It is a more robust technique and handles varieties of parts as compared to the syntactic method. However, the ambiguous representation and predefined rules needed for every conceivable feature make this approach overburden and inflexible [22]. There are also many disadvantages associated with the logic-based approach such as [4]

- The rules are nonunique to a feature
- These rules cannot be devised for every conceivable feature
- The recognition involves repeated exhaustive searches of the solid model

3.2.3 Graph-Based Approach

The graph-based approach is one of the important feature recognition methods that recognizes the features by matching the feature graph to the appropriate subgraph [1]. In the graph-based approach, a class of features is modeled using a graph structure, which presents the required topological and geometric constraints for identifying a given feature [3]. These graph structures can then be coded in various computational forms [3]. The graph-based approach was developed by Joshi in 1987 [23] in order to develop such a part representation in which a topological information and some geometric information of the part can be included. The B-rep model of the part (designed in some solid modeler) was transformed into the attributed adjacency graph (AAG). The AAG is a graph in which every arc takes attribute 0, if its nodes have a concave adjacency relation or 1, if they have a convex adjacency relation. The different steps involved in the graph-based feature recognition are as follows [9]:

- Generation of the graph-based representation of the part.
- Definition of the part features.
- Matching of the part features in the graph representation.

An AAG can be defined as a graph $G=(N, A, T)$, where N is the set of nodes, A is the set of arcs, and T is the set of attributes to arcs in A such that

- For every face f in F, there exists a unique node n in N.
- For every edge e in E, there exists a unique arc a in A, connecting nodes n_i and n_j, corresponding to face f_i and face f_j, which share the common edge e.
- The "t" is an attribute assigned to every arc a in A, where:
  - $t=0$ if the faces sharing the edge form a concave angle (or "inside" edge).
  - $t=1$ if the faces haring the edge form a convex angle (or "outside" edge).

The AAG is represented in the form of matrix as follows:

$$\begin{array}{c} \\ F_1 \\ F_2 \\ - \\ - \\ F_n \end{array} \begin{array}{c} F_1\,F_2 \ldots\ldots\ldots F_n \\ \begin{bmatrix} E_{1,1}\,E_{1,2}\ldots\ldots E_{1,n} \\ \ldots\ldots\ldots\ldots\ldots\ldots \\ \ldots\ldots\ldots\ldots\ldots\ldots \\ \\ E_{n,1}\,E_{n,2}\ldots\ldots\ldots E_{n,n} \end{bmatrix} \end{array}$$

$$E_{i,j} = \begin{cases} 0 \text{ if } F_i \text{ forms a concave angle with } F_j \\ 1 \text{ if } F_i \text{ forms a convex angle with } F_j \\ \Phi \text{ if } F_i \text{ is not adjacent to } F_j \end{cases}$$

The graph-based algorithms arrange the B-rep of a part into a graph structure. These graphs can have either faces, edges, or vertices as nodes, and any of the other two entities as arcs [4]. These graphs are split into subgraphs using a well-defined algorithm. The strength of the graph-based approaches lies in their ability to recognize the isolated features while they have shortcoming in recognizing the interacting features and multiple interpretations [1,24]. It requires extensive preprocessing to construct the graphs, and additional computation to extract the feature subgraph from the rest of the graph [4]. It is an efficient technique to recognize simple as well as complex features [2]. The advantages of graph-based recognition can be summarized as follows [1]:

- Applies to several domains, that is, it is not limited only to the machining
- Allows the users to add new feature types without changing the code
- Suitable for incremental feature modeling
- Recognize isolated features effectively

The AAG is constructed in such a way that every face becomes a unique node and every edge exists as a unique arc [8]. The feature recognition procedure is then called to compare the nodes of the graph to a predefined feature library. If the configuration of a given predefined feature matches that of the graph, a feature is identified and extracted out.

3.2.4 Expert System Approach

This approach is based on the transformation of the knowledge and the experience of the expert into the set of rules to solve the given problem. It consists of two primary elements [17,25]: the production knowledge and an inference engine. The production knowledge is made up of the procedural knowledge and the declarative knowledge. The procedural knowledge describes the production rules concerned with the generation of the process plan from the part design. It may include the feature sequence rules, raw material selection rules, tolerance rules, operation selection rules, machine tool selection rules, cutting tool selection rules, cutting parameters rules, and finishing operation rules. The production rules are generally expressed using If–Then statements such as if the given condition occurs, then do this action, otherwise, execute the other action. The following rule is an example of representing a slot. The user needs to input the required tolerances for machining this slot, and execute the slot milling rules [17]:

IF	the feature is a slot
AND	the slot dimensions are: *X*, *Y*, and *Z*
THEN	Input the slot tolerances
	Upper length tolerance
	Lower length tolerance
	Upper width tolerance
	Lower width tolerance
	Upper thickness tolerance
	Lower thickness tolerance
and	execute the slot milling rules.

The declarative knowledge is comprised of the production facts that are the feature facts, machine tool capacity facts, machining operation facts, and raw material facts [23,26]. The following rule represents the available feature facts in an expert process planning [17]:

- *Features*:
  - Rectangular slot
  - Cylindrical hole
  - Thread
  - Key
  - Pocket
  - Shaft
  - Step shaft

The inference engine of the expert process planning system can be defined as the control system of the executed facts and rules that is applied to solve a specific problem using searching algorithm [26]. The expert system to extract form features such as holes and slots may consist of three modules [22]:

- Feature recognition
- Feature extraction
- Feature graph construction

A Prolog format or any other language can be used to develop a regional shape pattern recognizer. The heuristic rules for defining the features are coded in Prolog, Lisp, or some other languages, which are known as the building tools for developing the expert systems.

3.2.5 Volume Decomposition and Composition Approach

This approach was basically introduced to generate the machining models [27,28]. The volume decomposition method [29,30] is based on decomposing the removal (machining) volume into the convex cells and then combining these cells to create machining features [1]. It actually decomposes the input (solid) model into a group of intermediate volumes and converts these volumes to produce features based on certain specified rules [7].

The difficulty with this type of approach is the conversion of the volumes into meaningful shapes of the part and the machining volumes [7]. The primary steps of the cell-based (or cellular) decomposition approach includes the following [7]:

- Determination of the overall removable volume by computing the difference between the blank and the finished part. In fact, the Boolean difference between the volume of the stock and the volume of the final part produces the total volume to be removed [8].
- Decomposition of this volume into unit volumes by utilizing the extended boundary faces as cutting planes (cell decomposition) or by extending and intersecting the surfaces or half space.
- Finally, merging of all the unit volumes sharing common faces or possessing coplanar faces to get maximum cells that can be removed in a single tool path (cell composition). The alternative machining features can also be generated by varying the composition sequence of the cells [9].

This method can successfully be applied to the feature interactions and the multiple interpretations of features [1]. It can be utilized for the polyhedral parts due to the complex convex hull computation of the curved objects. However, this technique is computationally complex, exhibits the inability to generate nonconvex delta machining features, and does not guarantee the generation of the correct set of machining features [1,31]. This method cannot be used directly to generate the machining features. Instead, they generate form features that are then converted into the machining features [32,33].

3.2.6 3D Feature Recognition from a 2D Feature Approach

This approach is used to recognize the machining features of the prismatic part comprising the planar or cylinder surfaces in terms of 2D profiles [34]. The 3D feature recognition using the 2D CAD data may include the following steps [35]:

- Extraction of the 2D geometric entities
- Divide and conquer strategy to extract the vertex-edge data from each 2D orthographic view (front, side, and top)
- Application of the production rules to recognize each 2D entity such as square, rectangle, triangle, or circular loop
- Finally, the 3D recognition can be performed

It also works by recognizing the 2D features from the polyhedral faces of the part model and then 3D feature is determined using the 2D feature [36]. The application of this procedure involves the determination of the nonconvex faces of the part model. Then, the rule-based and graph-based approach can be utilized to get the 3D feature using these extracted 2D features. The edges in the 2D feature are defined as the entities adjacent to the two faces (the primary and the secondary face). The face, which is associated with nonconvex faces in 2D recognition, is considered as the primary face. The other face, which is perpendicular, is considered to be the secondary face. This approach has the limit of recognizing only a limited number of features and also it does not identify the dimensions of the recognized features [37]. The features in this approach are limited to the prismatic features only and do not include the interaction between the features [35]. This approach can also be applied to the coordinate measuring machine (CMM) [38]. It includes the feature extraction using the IGES or the STEP file and then representing the entities in 2D using the orthogonal views (front, top, and side). Finally, the generation of the 3D shaped using the extracted feature.

3.3 Summary

The different feature recognition techniques have been elaborately explained in this chapter. The feature representation methods such as CSG, B-rep, and hybrid (CSG and B-rep) methods have also been discussed. This chapter also provides the benefits and the limitations of the feature recognition approaches including the syntactic pattern recognition, the logic-based, graph-based, expert system, volume decomposition, and composition and the 3D feature recognition from a 2D feature.

QUESTIONS

1. Define feature recognition. Write down different steps to accomplish the feature recognition process.
2. What is the significance of the feature recognition in the manufacturing industries?
3. Write down the difference between the human-assisted and the automatic feature recognition.
4. How is the feature recognition different from the design by feature technique?
5. What do you mean by the feature representation? What are the different techniques of feature representation?
6. What are the characteristics of the B-rep method?
7. Explain the difference between the constructive solid geometry (CSG) and boundary representation (B-rep).
8. What are the several advantages of the hybrid (B-rep and CSG) feature representation method?
9. Explain the syntactic pattern recognition approach. Write down its advantages and disadvantages.
10. Describe the slot and pocket feature using the logic-based approach.
11. How do you recognize the feature hole in logic-based approach using the CSG representation?
12. Write down the several benefits and the limitations of the logic-based approach.
13. What are the different steps involved in the graph-based approach?

14. Define the attributed adjacency graph (AAG) and describe the transformation of the B-rep model to the AAG.
15. Write down the various advantages and disadvantages of the graph-based approach.
16. What do you mean by the expert system approach? Explain the different components of the expert system.
17. Write down the different steps of the cell-based and the decomposition approach.
18. Write down the different applications of the volume decomposition and composition approach. What are its various obstacles?
19. Explain how a 3D feature recognition is different from a 2D feature approach.
20. What are the different steps involved in the 3D feature recognition using the 2D CAD data?

References

1. S. Gao and J. J. Shah, Automatic recognition of interacting machining features based on minimal condition subgraph, *Computer-Aided Design* 1998;30(9):727–739.
2. B. Venu, V. Komma, and D. Srivastava, A New Hybrid Approach to Recognize Machinable Features of Prismatic Parts from STEP AP 203 File, *5th International & 26th All India Manufacturing Technology, Design and Research Conference (AIMTDR 2014)*, December 12–14, 2014, IIT Guwahati, Assam, India.
3. M. C. Wu and C. R. Liu, Analysis on machined feature recognition techniques based on B-rep, *Computer-Aided Design* 1996;28(8):603–616.
4. S. Prabhakar and M. R. Henderson, Automatic form-feature recognition using neural-network-based techniques on boundary representations of solid models, *Computer-Aided Design* 1992;24(7):381–393.
5. H. C. W. Lau, C. K. M. Lee, B. Jiang, I. K. Hui, and K. F. Pun, Development of a computer-integrated system to support CAD to CAPP, *International Journal of Advances in Manufacturing Technology* 2005;26:1032–1042.

6. T. De Martino, B. Falcidieno, F. Giannini, S. Hassinger, and J. Ovtcharova, Feature-based modelling by integrating design and recognition approaches, *Computer-Aided Design* 1994;26(8):646–653.
7. B. Babic, N. Nesic, and Z. Miljkovi, A review of automated feature recognition with rule-based pattern recognition, *Computers in Industry*, 2008;59:321–337.
8. M. P. Bhandarkar and R. Nagi, STEP-based feature extraction from STEP geometry for agile manufacturing, *Computers in Industry* 2000;41:3–24.
9. J. Gao, D. T. Zheng, and N. Gindy, Extraction of machining features for CAD/CAM integration, *International Journal of Advances in Manufacturing Technology* 2004;24:573–581.
10. S. H. Chuang and M. R. Henderson, Three-dimensional shape pattern recognition using vertex classification and vertex-edge graphs, *Computer-Aided Design* 1990;22(6):377–387.
11. J. J. Shah, Assessment of features technology, *Computer-Aided Design* 1991;23(5):331–343.
12. S.-C. Liu, M. Gonzalez, J.-G. Chen, Development of an automatic part feature extraction and classification system taking CAD data as input, *Computers in Industry* 1996;29:137–150.
13. U. Roy and C. R. Liu, Feature-based representational scheme of a solid modeler for providing dimension and tolerancing information, *Robotics and Computer-Integrated Manufacturing* 1988;4(3/4):335–354.
14. N. Wang and T. M. Ozsoy, A scheme to represent features, dimensions, and tolerances in geometric modeling, *Journal of Manufacturing System*, 1991;10(3):233–240.
15. R. Jakubowski, Syntactic characterization of machine-parts shapes, *Cybernetics and Systems Integration Journal* 1982;13:1–24.
16. S. Staley, M. Henderson, and D. Anderson, Using syntactic pattern recognition to extract feature information from a solid geometric model data base, *Computers in Mechanical Engineering* 1983;61–66.
17. E. A. Nasr and A. K. Kamrani, *Computer-Based Design and Manufacturing: An Information-Based Approach*, Berlin: Springer, 2007.
18. V. B. Sunil and S. S. Pande, Automatic recognition of features from freeform surface CAD models, *Computer-Aided Design* 2008;40:502–517.

19. B. K. Choi, N. M. Barash, and D. C. Anderson, Automatic recognition of machined surfaces from a 3-D solid model *Computer-Aided Design* 1984;16(2):81–86.
20. M. R. Henderson, Extraction of feature information from three dimensional CAD data, PhD thesis, Purdue University, IN, May 1984.
21. M. R. Henderson and D. C. Anderson, Computer recognition and extraction of form features: A CAD/CAM link, *Computers in Industry* 1984;5:329–339.
22. M. R. Henderson and D. C. Anderson, Computer recognition and extraction of form features: A CAD/CAPP link, *Computers in Industry* 1984;5:329–339.
23. S. Joshi, T. C. Chang, Graph based heuristics for recognition of machined features from 3D solid model, *Computer Aided Design* 1998;20(2):58–66.
24. K. Rahmani and B. Arezoo, Boundary analysis and geometric completion for recognition of interacting machining features, *Computer-Aided Design* 2006;38:845–856.
25. G. F. Luger and W.A. Stubblefield, *Artificial Intelligence and the Design of Expert Systems*, Menlo Park, CA: Benjamin/Cummings, 1989.
26. H. P. Wang and R. A. Wysk, An expert system of machining data selection, *Computer and Industrial Engineering Journal* 1986;10(2):99–107.
27. Y. Tseng and S. B. Joshi, Recognising multiple interpresentations of interacting machining features, *Computer-Aided Design* 1994;26(9):667–688.
28. H. Sakurai and C. Chin, Definition and recognition of volume features for process planning, *Advances in Feature-Based Manufacturing* 1994;1994:65–80.
29. Y. Kim, Recognition of form features using convex decomposition, *Computer-Aided Design* 1992;24(9):461–476.
30. Y. Kim and E. Wang, Recognition of machining features for cast then machined parts, *Computer-Aided Design* 2002;34:71–87.
31. J. H. Han, M. Pratt, and W. C. Regli, Manufacturing feature recognition from solid models: A status report, *IEEE Transactions on Robot Automation* 2000;16(6):782–796.
32. H. Sakurai and C. Chin, Defining and recognizing cavity and protrusion by volumes. *Proceedings of the ASME Computers in Engineering Conference*, 59–65, 1993.

33. Y. Woo and S. Sakurai, Recognition of maximal features by volume decomposition, *Computer-Aided Design* 2002;34:195–207.
34. S. Meeran and M. J. Pratt, Automatic feature recognition from 2D drawings, *Computer-Aided Design* 1993;25(1):7–17.
35. C.-H. Liu, D.-B. Perng, and Z. Chen, Automatic form feature recognition and 3D part recognition from 2D CAD data, *Computer and Industrial Engineering* 1994;14(4):689–707.
36. V. Allada and S. Anand, Machine understanding of manufacturing features, *International Journal of Production Research* 1996;34(7):1791–1820.
37. M. G. L. Sommerville, D. E. R. Cleark, and J. R. Corney, Viewer-centered feature recognition, *Proceedings: Symposium on Solid Modeling and Applications*, pp. 125–129, 1995.
38. G. X. Zhang, S. G. Liu, X. H. Ma, and Y. Q. Wang, Toward the intelligent CMM, *CIRP Annuals, Manufacturing Technology* 2002;51(1):437–442.

Chapter 4

Data Transfer in CAD/ CAM Systems

4.1 Data Transfer in CAD/CAM Systems

A large volume of technical information and data about the products has to be negotiated (or transferred) between various systems in the design and manufacturing industries. The different systems have their own distinct data format; therefore, a given set of information has to pass through various data formats and the different systems, thus, leading to increased errors or missing information. However, this problem of data transfer between various systems has been solved to a very large extent, thus, enabling reliable data transfer across various systems.

Data translation is nothing but a process of conversion, where, the information from one system (with its own data format) is converted to the format compatible to the other concerned system. For example, a given design file has to be exported and imported through various CAD systems before the final is made available for manufacturing. As shown in Figure 4.1, the data translation process commences with a CAD model exported from CATIA as STEP A203. This file

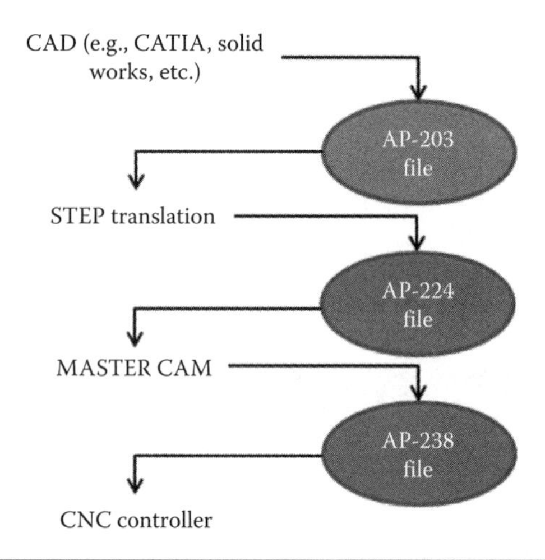

Figure 4.1 Data translation process. (Adapted from S. Ranđelović and S. Živanović, *Mechanical Engineering,* 2007;5(1):87–96.)

(STEP A203) is translated to produce an AP224 data file, which can be read by MASTERCAM. Finally, MASTERCAM generates a AP238 file containing the information related to the geometry, feature, and tool path for the machine controller [1].

There are two different types of translators for data transfer: direct translators and neutral translators, as shown in Figure 4.2. Direct translators convert the data or the information in one CAD/CAM system (or format) to the format of another CAD/CAM system. They are dedicated translators which link the two specific CAD/CAM packages [2]. They ensure a reliable and complete exchange of information between the two concerned systems [3]. The disadvantage with this translator is that the complexity increases as the number of communicating systems increases. For N systems, $N(N-1)$ direct translators are required while $2(N-1)$ of these will require change with change in any system [3]. On the other hand, the neutral translators convert the data or information into a neutral file (such as IGES, STEP, etc.), which is independent of the format of the other CAD/CAM system. With a neutral file, only two translators are required for each system. Therefore, for N systems, $2N$ translators are required. If one

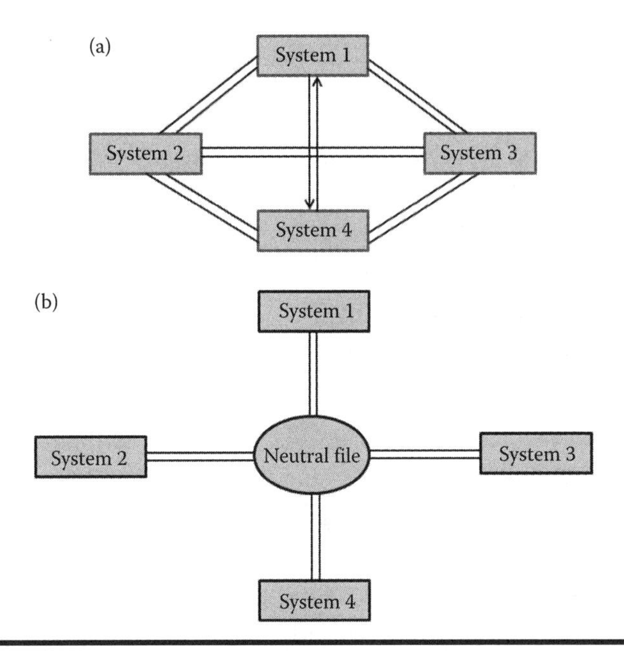

Figure 4.2 (a) Direct translator and (b) neutral translator.

system is changed, only two translators associated with it are modified [3]. With neutral format, there are chances of data transfer errors, which, in turn, increases the product's manufacturing cycle. Direct translators are faster than the neutral translators and the size of the file generated by direct translators is smaller than the one produced by the neutral translators.

The translator associated with a particular system is called as preprocessor as shown in Figure 4.3. It converts the internal system (CAD/CAM) data to the neutral file format while the postprocessor transforms the neutral file format to the given CAD/AM system [3].

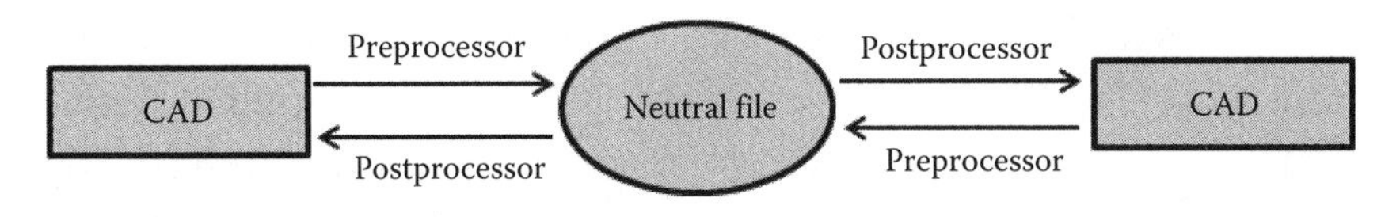

Figure 4.3 Representation of preprocessor and postprocessor. (Adapted from E. A. Nasr and Ali K. Kamrani, *Computer-Based Design and Manufacturing: An Information-Based Approach*, Berlin: Springer, 2007.)

4.1.1 Initial Graphics Exchange Specifications

Initial graphics exchange specification (IGES) was first developed by the National Aeronautical and Space Administration and the National Bureau of Standards in 1979. It has been adapted and recognized by the American National Standard Institute (ANSI) as a standard tool format. Consequently, IGES has become an acceptable and widely used neutral format translator by many CAD/CAM system vendors [5]. Although some translators have been more popular than IGES, this neutral format translator has been through many revisions and has proved to be a comprehensive tool in transferring data for parts designed by wireframe, surface, or solid models.

The IGES format serves as a neutral data format to transfer the design to a dissimilar system [6]. Translators, developed to the IGES Standard, are used to export a design into an IGES file for exchange and for importing the IGES file into the destination system. An IGES file is a sequential file incorporating a sequence of records [4]. The fundamental unit of data in the IGES file is the entity. The IGES file is written using the ASCII characters as a sequence of 80 character records. The entities can be categorized as geometric, annotation, and structure entities [4]. The geometric entities represent the definition of the physical shape, which include points, arcs, curves, surfaces, solids, etc. The annotation entities comprise dimensions, notes, sketching plane, etc., which help in the documentation and visualization of the object. Finally, the structure entities specify the associations between other entities in the IGES file. An IGES file consists of six sections in the order of flag section, start section, global section, directory entry (DE) section, parameter data (PD) section, and terminate section, as shown in Figure 4.4 [7,8].

1. Section 1: FLAG section—This section is optional and it indicates the structure in which the data is specified. For example, ASCII format. Since ASCII format results in very

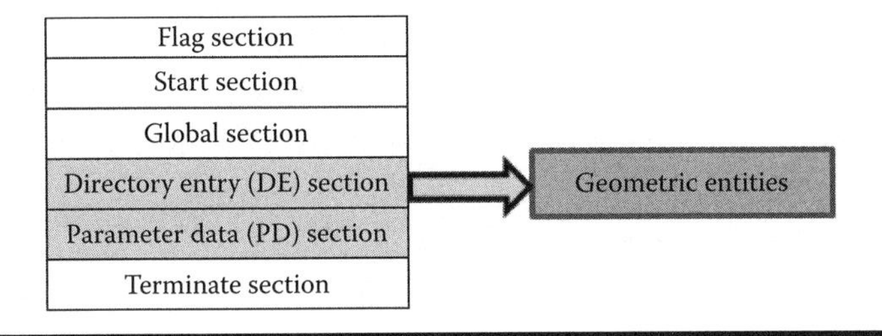

Figure 4.4 Structure of IGES file.

large file sizes, the format can be standardized in either of the following three modes [8]:

a. ASCII
b. Binary form
c. Compressed ASCII

The binary form and compressed ASCII aid in reducing the size of the file.

2. Section 2: START section—It is, in fact, a human readable introduction of the file. It can be defined as a "prologue" to the IGES file. It provides information such as sending source (name of the CAD/CAM system), receiving station (CAD/CAM system), description of the product being converted, among others. This information is of great help to the personnel who are postprocessing the file for the downstream applications.
3. Section 3: GLOBAL section—It consists of the information about the product, name of the IGSE file, the author of the file, the source (sending) company (or the vendor), software version of the source system, the date and time of the file generation, drafting standards, model units, minimum resolution, and maximum coordinate values. It also provides information about the preprocessor and the postprocessor in order to interpret the file on the destination computer.
4. Section 4: DIRECTORY ENTRY (DE) section—It consists of the list of all the entities that have been defined in the IGES file along with the attributes associated with them.

It provides an index of the file and contains attribute information such as color, line type, and transformation matrix. It also comprises the pointer to the parameter data section entry, which contains the required parameter data. The DE provides an index, organized in fixed fields, and are consistent for all entities to provide simple access to frequently used descriptive data.

5. Section 5: PARAMETER DATA (PD) section—This section contains the data associated with the entities listed in the DE section. For example, a straight line entity can be defined by the six coordinates of its two endpoints. The parameter data provides the specific entity definition and is entity-specific and variable in length and format.
   The DE data and PD for all entities in the file are arranged into separate sections, with pointers providing bidirectional links between the DE and PD for each entity.
6. Section 6: TERMINATE section—The terminate section consists of a single record. This record details about the number of records in each of the preceding sections for verification purposes.
   The main problem with the IGES files is their large size, which ultimately leads to the higher processing time. They are also prone to errors during pre- and postprocessor implementations.

4.1.2 Standard for Exchange of Product Data

STandard for the Exchange of Product model data [STEP (ISO 10303)] represents a series of international standards for defining and exchanging product model information across an engineering and manufacturing lifecycle [7]. It has been supported by most of the major CAD software. The data specification language used to represent the product information in this format is EXPRESS. There are a number of benefits that can be associated with the implementation of STEP:

- It can transfer simple solids and surface geometry.
- It is the most robust neutral format for 3D solid and surface geometry.
- Neutral file types such as IGES generate a large size file for complex geometries as compared to STEP as shown in Figure 4.5.
- STEP translators are available for most of the major CAD/CAM systems.
- STEP does not only define the geometric shape of a product but also includes topology, features, tolerance specifications, material properties, etc. It provides all necessary information required to completely define the product for the design, analysis, manufacture, test, inspection, and product.
- It has been an evolving standard that will also include the aspects such as data sharing, storage, and exchange.

The primary objective for the development of STEP is to define a standard that incorporates all facets of the product such as geometry, topology, tolerances, materials, etc. In STEP, the representations of the product information and implementation methods are separated. The implementation methods are used for data exchange while representation provides the definition of product information to various applications.

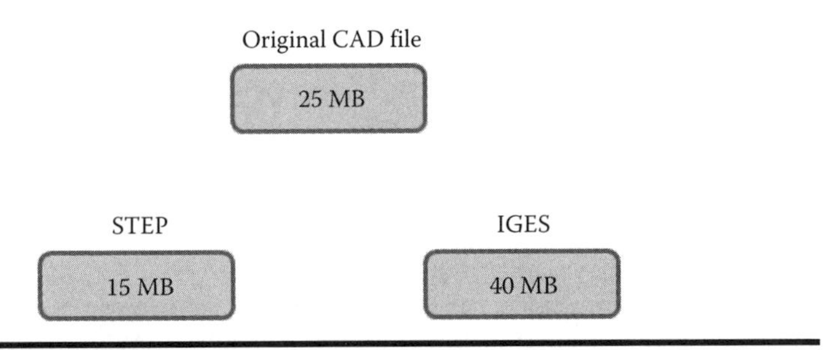

Figure 4.5 Size comparison of STEP and IGES files. (Adapted from A. Papavasileiou et al., CAD/CAM—Interfaces: A Review. *2002 First International IEEE Symposium "Intelligent Systems" Proceedings*, vol. 3, pp. 378–382, September 2002.)

4.1.2.1 Structure of STEP

STEP can be divided into a number of modules such as description methods, integrated resources, application protocols, abstract test suites, implementation methods, and conformance testing [9] as shown in Figure 4.6. The structural components and functional aspects of STEP can be described as follows [11]:

- 0 Series: Introduction
- 10 Series: Description methods

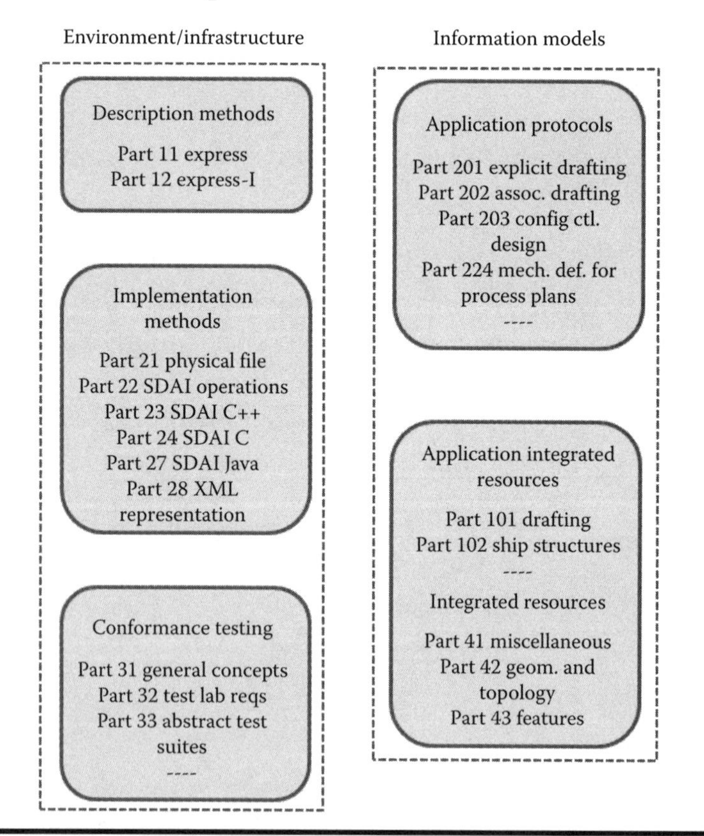

Figure 4.6 Structure of STEP. (Adapted from S. M. Amaitik, 2005, Developing of a STEP feature-based intelligent process planning system for prismatic parts, Doctoral Dissertation, Middle East Technical University; Steptools, http://www.steptools.com/library/fundimpl.pdf [accessed on June 8, 2015].)

- 20 Series: Implementation methods
- 30 Series: Conformance testing methodology and framework
- 40 Series: Integrated generic resources
- 100 Series: Integrated application resources
- 200 Series: Application protocols (APs)
- 300 Series: Abstract test suites
- 500 Series: Application interpreted constructs

4.1.2.1.1 Introductory Documents

This series explains the objective and structure of STEP. It also provides an explanation of some basic terms. It forms the part one of ISO 10303, which provides overview and fundamental principles of STEP.

4.1.2.1.2 Description Methods

The description method describes the major component of the series of STEP modules. It explains the languages and methods that are used to generate the STEP file. The description methods are standardized in the ISO 10303-10 series of parts. The different series of description method can be given as follows:

- ISO 10303-11: EXPRESS language reference manual
- ISO 10303-12: EXPRESS-I language reference manual
- ISO 10303-13: Architecture and methodology reference manual
- ISO 10303-xx: EXPRESS-M language reference manual

The EXPRESS language is used to represent the structure of the data and the integrated resources and application protocols are also defined using EXPRESS.

4.1.2.1.3 Implementation Methods

It explains the communication between the STEP and other formal languages. It in fact includes the methods of representing data that has been modeled in EXPRESS. This series includes the physical file exchange structure, the standard data access interface, and its language bindings [9]. It provides the specifications of how the information in STEP file can be represented for the exchange environment [7]. The implementation methods are standardized in the ISO 10303-20 series of parts.

- ISO 10303-21: Clear text encoding of the exchange structure
- ISO 10303-22: Standard data access interface (SDAI)
- ISO 10303-23: C++ language binding to SDAI
- ISO 10303-24: C language binding to SDAI
- ISO 10303-25: FORTRAN language binding to SDAI
- ISO 10303-28: XML representations of EXPRESS schemas and data

4.1.2.1.4 Conformity Tests

It examines the degree of conformity of the software associated with the application protocol. It provides the concepts of conformance testing as well as actual test methods and requirements on testing labs and clients. It provides following information [7]:

- Instructions for developing abstract test suites
- Techniques for testing of software—product conformance to the STEP standard
- Specifications for conformance testing of the processors
- Responsibilities of the testing laboratories

The conformance testing methods are standardized in the ISO 10303-30 series of parts [9].

- ISO 10303-31: General concepts
- ISO 10303-32: Requirements on testing laboratories and clients
- ISO 10303-33: Abstract test suites

- ISO 10303-34: Abstract test methods for part 21
- ISO 10303-35: Abstract test methods for part 22

4.1.2.1.5 Integrated Generic Resources

They represent the conceptual building blocks for the STEP file. The specifications of the information models such as geometric and topological representation, product structure organization, materials, tolerances, form features, and properties are contained in this series.

- ISO 10303-41: Fundamentals of product description and support
- ISO 10303-42: Geometric and topological representation
- ISO 10303-43: Representation structures
- ISO 10303-44: Product structure configuration
- ISO 10303-45: Materials
- ISO 10303-46: Visual presentation
- ISO 10303-47: Shape variation tolerance
- ISO 10303-48: Form features
- ISO 10303-49: Process structure and properties

4.1.2.1.6 Abstract Test Suites

It contains the set of abstract test cases necessary for conformance testing of an implementation of a STEP application protocol. Each abstract test case specifies input data to be provided to the implementation under test, along with information on how to assess the capabilities of the implementation.

- ISO 10303-301: Abstract test suite: Explicit draughting
- ISO 10303-302: Abstract test suite: Associative draughting
- ISO 10303-303: Abstract test suite: Configuration-controlled design
- ISO 10303-304: Abstract test suite: Mechanical design using boundary
- ISO 10303-305: Abstract test suite: Mechanical design using surface representation

- ISO 10303-307: Abstract test suite: Sheet metal die planning and design
- ISO 10303-308: Abstract test suite: Lifecycle product change process

4.1.2.1.7 Application Protocols

They represent the central component of the STEP architecture, and the STEP architecture is designed primarily to support and facilitate development of APs. These are the main protocols that are used for the exchange of data between specific application systems. They don't only describe what data is to be used in describing the product, but also how the product is to be used in the model.

- ISO 10303-201: Explicit draughting
- ISO 10303-202: Associative draughting
- ISO 10303-203: Configuration-controlled design
- ISO 10303-204: Mechanical design using boundary representation
- ISO 10303-205: Mechanical design using surface representation
- ISO 10303-206: Mechanical design using wireframe representation
- ISO 10303-207: Sheet metal die planning and design
- ISO 10303-208: Lifecycle product change process
- ISO 10303-209: Design analysis of composite structures
- ISO 10303-210: Printed circuit assembly product design data
- ISO 10303-211: Electronic test, diagnostics, and remanufacture
- ISO 10303-212: Electrotechnical design and installation
- ISO 10303-213: Numerical control process plans for machined parts
- ISO 10303-214: Core data for automotive mechanical design processes
- ISO 10303-215: Ship arrangements
- ISO 10303-216: Ship molded forms

- ISO 10303-217: Ship piping
- ISO 10303-218: Ship structures
- ISO 10303-219: Dimensional inspection process planning
- ISO 10303-220: Printed circuit assembly manufacturing planning
- ISO 10303-221: Functional data and schematic representations of process plants
- ISO 10303-222: Design-manufacturing for composite structures
- ISO 10303-223: Exchange of design and manufacturing information for cast parts
- ISO 10303-224: Mechanical product definition for process planning using the form
- ISO 10303-226: Ship mechanical systems
- ISO 10303-227: Plant spatial configuration
- ISO 10303-228: Building services: Heating, ventilation and air condition
- ISO 10303-230: Building structural frame: Steelwork

4.1.2.1.8 Application Interpreted Constructs

The application interpreted construct defines the data structures and semantics that are utilized to exchange product data common to two or more application protocols [9]. It represents the specific resources that are useful to express the identical semantics in more than one application protocol. It includes edge-based wireframe, draughting elements, shell-based wireframe, constructive solid geometry, geometry-bounded 2D wireframe, drawing structure, manifold surfaces, and nonmanifold surfaces [7].

- ISO 10303-502: Topology bounded surface and Breps ISO 10303-506: Draughting elements
- ISO 10303-507: Geometrically bounded surface
- ISO 10303-509: Manifold surface
- ISO 10303-510: Geometrically bounded wireframe
- ISO 10303-512: Faceted Brep

- ISO 10303-513: Elementary Brep
- ISO 10303-514: Advanced Brep
- ISO 10303-515: Constructive solid geometry
- ISO 10303-516: Mechanical design context

There are two primary sections in the STEP file: header and data section [12] as shown in Figure 4.7. The header section consists of a file description, file name, author,

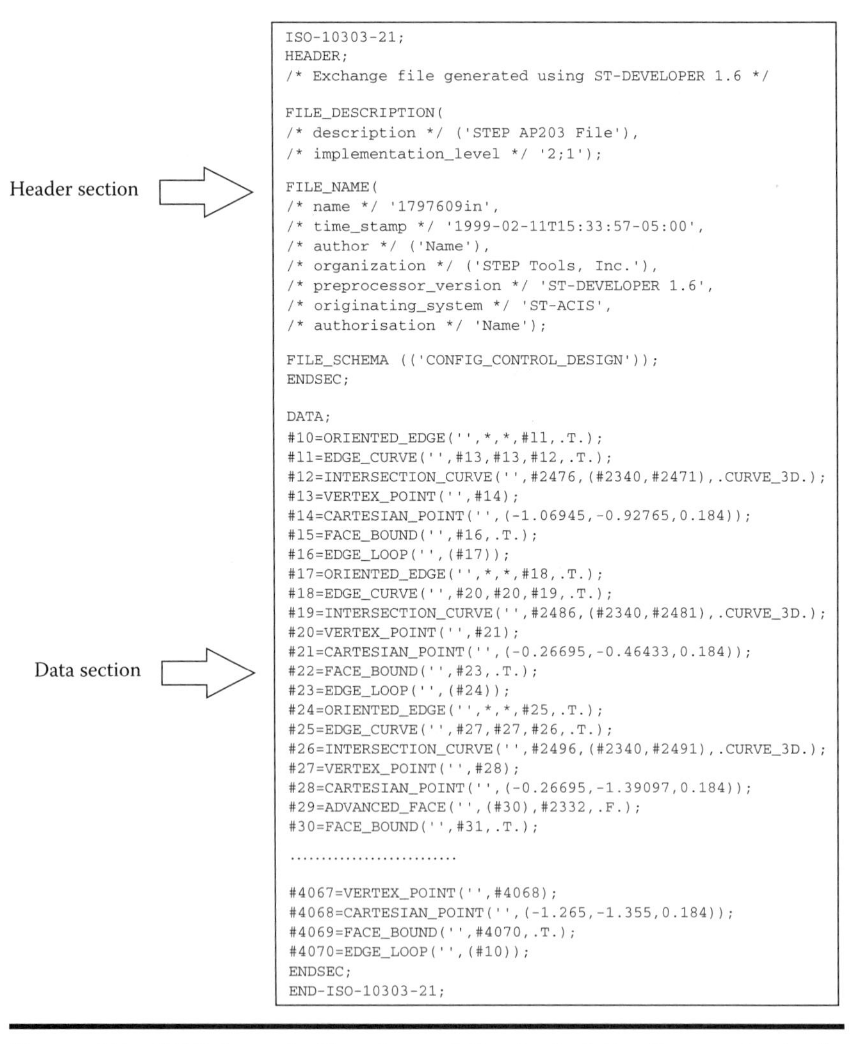


```
ISO-10303-21;
HEADER;
/* Exchange file generated using ST-DEVELOPER 1.6 */

FILE_DESCRIPTION(
/* description */ ('STEP AP203 File'),
/* implementation_level */ '2;1');

FILE_NAME(
/* name */ '1797609in',
/* time_stamp */ '1999-02-11T15:33:57-05:00',
/* author */ ('Name'),
/* organization */ ('STEP Tools, Inc.'),
/* preprocessor_version */ 'ST-DEVELOPER 1.6',
/* originating_system */ 'ST-ACIS',
/* authorisation */ 'Name');

FILE_SCHEMA (('CONFIG_CONTROL_DESIGN'));
ENDSEC;

DATA;
#10=ORIENTED_EDGE('',*,*,#11,.T.);
#11=EDGE_CURVE('',#13,#13,#12,.T.);
#12=INTERSECTION_CURVE('',#2476,(#2340,#2471),.CURVE_3D.);
#13=VERTEX_POINT('',#14);
#14=CARTESIAN_POINT('',(-1.06945,-0.92765,0.184));
#15=FACE_BOUND('',#16,.T.);
#16=EDGE_LOOP('',(#17));
#17=ORIENTED_EDGE('',*,*,#18,.T.);
#18=EDGE_CURVE('',#20,#20,#19,.T.);
#19=INTERSECTION_CURVE('',#2486,(#2340,#2481),.CURVE_3D.);
#20=VERTEX_POINT('',#21);
#21=CARTESIAN_POINT('',(-0.26695,-0.46433,0.184));
#22=FACE_BOUND('',#23,.T.);
#23=EDGE_LOOP('',(#24));
#24=ORIENTED_EDGE('',*,*,#25,.T.);
#25=EDGE_CURVE('',#27,#27,#26,.T.);
#26=INTERSECTION_CURVE('',#2496,(#2340,#2491),.CURVE_3D.);
#27=VERTEX_POINT('',#28);
#28=CARTESIAN_POINT('',(-0.26695,-1.39097,0.184));
#29=ADVANCED_FACE('',(#30),#2332,.F.);
#30=FACE_BOUND('',#31,.T.);
..........................

#4067=VERTEX_POINT('',#4068);
#4068=CARTESIAN_POINT('',(-1.265,-1.355,0.184));
#4069=FACE_BOUND('',#4070,.T.);
#4070=EDGE_LOOP('',(#10));
ENDSEC;
END-ISO-10303-21;
```

Figure 4.7 STEP format. (Courtesy of Steptools, http://www.steptools.com/ [accessed on June 9, 2015].)

organization name, properties of instances, sending source, and receiving system. On the other hand, the data section comprises of instances that are specific to the currently employed application protocol (AP). Many APs have been developed so far to look various engineering applications and manufacturing processes [26]. Presently, the most widely used AP is the IS AP203 while AP201, AP224, and AP202 are also very common.

The IGES file format was initially developed for basic drawing entities like line, poly lines, etc. However, its later versions also included a B-rep and CSG. The IGES file format is good for exporting 2D CAD drawings and 3D surface models. The 3D solid models are exported as surfaces and then need to be healed and/or closed when imported. STEP file format, on the other hand, is good for 3D solid models. The STEP file format can handle both geometric as well as nongeometric data.

4.2 Dimensional Measuring Interface Standard

Dimensional measuring interface standard (DMIS) was developed by Advanced Manufacturing International for Coordinate Measuring Machines (CMMs). DMIS can be defined as a programming language (ASCII format), which allows the execution of inspection programs between different measuring devices. It is a programming language used for programming mechanical, optical, laser, and video measuring systems. It enables smooth and seamless data transfer between CMMs of various manufacturers. Earlier, each of the CMM companies used to develop their own programming language for measurement purposes. As a result, it was difficult for different manufacturers to execute each other CMM inspection part program [14]. In fact, the development of the DMIS standard has allowed the industries to use an inspection part program on any given CMM system. The primary purpose for the development of DMIS program was to provide a platform through which CMM programs can be transferred from any CAD/CAM environment to any CMM (independent of the manufacturer). It provides a bidirectional communication between the CAD/CAM systems and the CMM. The remarkable features of DMIS are that they enable standardization of a part program, provide a universal programming language irrespective of the measurement device, and facilitate offline part programming [14]. Moreover, the ease of use, the editing capability of the inspection program, and the ease of training the operators with the implementation of DMIS cannot be disregarded [15]. It has so far figured out the problem of inspection program translation to the built-in CMM language. DMIS also consists of error correction, which, in turn, enhances the measurement accuracy.

The inspection planning generated by computer-aided inspection planning (CAIP) can be exported to CMM by using dimensional measuring interface standard (DMIS). DMIS files must be postprocessed (or translated) so that they can be used by measuring machines. The translator tailored to relevant measuring software is referred to as the post processor. The entire information needed for the inspection part program can be combined in a device-independent DMIS format to convert it into an equipment-specific program.

A large collection of research work can also be pointed out here, where DMIS has been successfully implemented. For example, Lin and Murugappan [16] proposed the integration of CMM into the CAD/CAM environment for generating the inspection path using CAD. The applied algorithm generated the inspection path using the CAD/CAM system and then extended it to a feature-based inspection planning. The tolerance specifications were also incorporated, and a DMIS output was created for online analysis. Similarly, Hermann [17] presented an offline programming system for CMM. The system used the CAD model as an input and the user could select the required surface elements for inspection. The system automatically generated the optimal distribution of the measuring points for the surfaces to be inspected and the local probe path. All the information needed for the CMM program was combined in a device-independent DMIS format and then converted to equipment-specific programs. Moreover, the automated system proposed by Sathi and Rao [18] used the STEP (ISO 10303) file of the part to be inspected as an input and provided the inspection process plan as an output in DMIS (ISO 22903) format. A DMIS program was also developed by Yuewei et al. [19] to execute the measuring tasks accurately and efficiently. This program saved about 30% time as compared to conventional programming method.

4.2.1 Components of DMIS File

DMIS is a standard format of high-level programming language, which transfers the equipment and product definitions along with the process and recording the information necessary to perform G&DT measurements that employ coordinate metrology. DMIS contains definitions of the nominal features of the product, dimensional, and geometric tolerances, feature constructions, part coordinate systems, and functional data. It communicates equipment definitions for various measurement resources, machine parameters, and measurement sensors. DMIS instructs touch probing and sensor's motions for manufacturing process validation and control [20]. It is used for bidirectional transfer of inspection data between CAD systems and CMM. The generated program of measuring feature is executed to compare GD&T like position, orientation, size, flatness, etc., depending on allowable tolerances associated with feature [21].

- **Base Alignment:** Datum alignment or part setup, as it is sometimes called, is required in order to set the position and orientation of CMM coordinate system. The part setup is usually defined by geometric tolerance definitions contained in component drawings.

- **Sensor Procedure: Sensor definition**
  - **SNSDEF** used for sensor definition.
    - S(name) = SNSDEF/PROBE, type, CART, x, y, z, i, j, k, length, diameter
- Selection of previously defined sensors
  - **SNSLCT/SA(name)**
    - An SNSLCT statement is used to select a sensor with the SA option sensor names.
- **Calibration sensor**
  - **CALIB/SENS, SA(name), FA(name), n**
    - CALIB is a command used for calibrating the previously defined sensor.
    - S(name) is the name of the sensor used to calibrate.
    - FA(name) is the name of a previously defined or measured feature that represents the calibration master artifact.
- **Feature Definition**
  - F(name) = FEAT/type, Cartesian, direction

    FEAT defined the name, coordinate information, position and the normal vector direction of the feature. The coordinate information may be a Cartesian (rectangular) three-dimensional position (x, y, z), or a polar (circular) two dimensional position [21].
    - Arc definition using start and finish angles

      F(name) = FEAT/ARC, dir, coordinate, x, y, z, i, j, k, radius, a1, a2
    - Circle definition

      F(name) = FEAT/CIRCLE, dir, coordinate, x, y, z, i, j, k, diameter
    - Slot definition

      F(name) = FEAT/CPARLN, dir, type, coordinate x, y, z, i1 ,j1, k1, i2, j2, k2 length, width
    - Cone definition

      F(name) = FEAT/CONE, dir, coordinate , x, y, z, i, j, k, angle

- Cylinder definition
  F(name) = FEAT/CYLINDER, dir, coordinate , x, y, z, i, j, k, diameter (,length)
- Curve definition
  F(name) = FEAT/GCURVE, coordinate , x, y, z, i, j, k
- Surface definition
  F(name) = FEAT/GSURF
- Line definition using vectors
  F(name) = FEAT/LINE, UNBND, coordinate , x, y, z, i1 ,j1, k1, i2, j2, k2
- Line definition using points
  F(name) = FEAT/LINE, BND, coordinate , x1, y1, z1, x2 ,y2, z2, i2, j2, k2
- Plane definition
  F(name) = FEAT/PLANE, coordinate , x, y, z, i, j, k
- Point definition
  F(name) = FEAT/POINT, coordinate , x, y, z, i, j, k
- Sphere definition
  F(name) = FEAT/SPHARE, dir, coordinate , x, y, z (,i, j, k)

■ Feature Measuring
  - MEAS/Feat _type, F(name), n
  - MEAS statement is used to measure the previously defined feature with MEAS-ENDMES measurement block
  - Feat _type: define the feature type
  - F(name): Defines name of this feature in the previous definition, and n number of touch points

■ Machine Movement: A GOTO is a statement used to move CMM automatically from one measuring feature to another measuring feature in a way that avoids probe collision [21]
  - Cartesian
  - GOTO/ x, y, z
  - Polar
  - GOTO/POL, radius, angle, height

  - Incremental
  - GOTO/INCR, distance, i, j, k
  - Defining a Cartesian home position
  - FORM/x, y, z
  - Defining a polar home position
  - FORM/POL, radius, angle, height
  - Moving to the home position
  - GOHOME
- Test of Geometrical and Dimensional Tolerance
  - Geometrical dimensional
    - It contains an upper limit and lower limit of dimension and acceptance depends on tolerance value
    - T(name) = TOL/size_type, tolerance-value, direction
    - Size_type: the dimension type required to measure like, length, width and height of the feature
- Form Tolerance: It is tolerance of shape verification depending on datum face selected
  - Position tolerance:
    T(name) = TOL/POS, tolerance-value
  - Perpendicularity tolerance:
    T(name) = TOL/PERP, tolerance-value, FA(datum)
  - Parallelism tolerance:
    T(name) = TOL/PARLEL, tolerance-value, FA(datum)
  - Angularity tolerance:
    T(name) = TOL/ANGLR, tolerance-value, FA(datum)
  - Concentricity tolerance:
    T(name) = TOL/CONCEN, tolerance-value, FA(datum)
  - Flatness tolerance:
    T(name) = TOL/FLAT, tolerance-value
  - Straightness tolerance:
    T(name) = TOL/ STRGHT, tolerance-value
- Result Output: It describes how to associate tolerances with previously measured or constructed features in order to create reports

- OUTPUT/FA(name), TA(name)
- FA(name): Represents the name of the actual feature needed to be reported
- TA(name): Represents the predefined result tolerance used to report the feature(s)

4.3 Object-Oriented Programming

Object-oriented programming (OOP) works by dividing a given problem into subgroups of related parts [22]. The different concepts exist in OOP comprises of objects, classes, data abstraction, inheritance, data encapsulation, and polymorphism.

The objects present a correspondence with the real-world object and can be defined as the primary run-time entities that are created in OOP. For example, it can represent a person, a bank account, or data table. The object maintains its state in one or more variables. The object can also be represented by a user defined data, such as vectors, time, lists, etc. The object takes up space in memory and has an associated address. Each object contains data and a code to manipulate the data.

A class is used to manufacture or create the objects. It can have many objects of similar kinds, such as production, employee, students, and so on. The class represents a group of objects with similar characteristics, identical properties, and shared relationship. [23]. After defining a class, any number of objects can be created from that class. For example, if a vehicle has been defined as a class, then the statement vehicle truck will create an object truck belonging to the class vehicle. The difference between classes and objects is that the objects are created and deleted with run-time program, and if many objects are created from the same class, then they must be in the same structure [24].

Data abstraction can be defined as a technique or the process of representing the essential information without background details and explanations [22]. It provides a distinct separation between the abstract properties of a data type and the specific details of its implementation [25]. Let's take a real world example of a mobile. The mobile can be used for different functions such as calling, watching videos, listening to music, and various additional functions including HDMI and USB can be used. However, the users do not know the internal details of the mobile. For example, how the mobile is receiving the signals, how the video is played, how it performs calling, etc. Therefore, it can be concluded from this example that a mobile explicitly separates its internal implementation with its external interface. In programming, the user can call a sort function without actually knowing what algorithm the function actually uses to sort the given values.

Encapsulation can be defined as a process of wrapping up of data and functions into a single component (class) [23]. For example, the data within a given class can only be accessed by the functions wrapped within this class. Due to encapsulation, data within a given class is inaccessible to the outside world.

Inheritance can be defined as a technique through which the objects of one class acquire the properties of the objects of another class [23]. For example, BMW X3 is a part of the class four-wheeler, which is also a part of the class vehicle. In fact, each of the derived class shares common characteristics with the class from which it has been derived.

Polymorphism can be defined as the ability of the function (or the operation) to exhibit different properties on different classes. For example, the operation function will generate a sum for two numbers, however, for strings, the same operation (sum) will generate a third string by concatenation [22].

The object-oriented programming language (such as C++) due to its unique characteristics and benefits provides a user-friendly platform for the integration of CAD/CAM systems and CMM systems.

4.4 Summary

Standardization of CAD/CAM information has become an important aspect of design and manufacturing industries. This is due to the fact that it enables smooth and error free exchange of information between various CAD/CAM systems. There exists two types of translators, such as direct and neutral translators for data exchange. However, the neutral translators such as IGES and STEP are the most preferred choices by the industrial personnel. Moreover, the standard such as DMIS has also been introduced to aid the engineers in exchanging information between CAD/CAM systems and CMM.

QUESTIONS

1. What do you mean by standardization of data in CAD/CAM systems?
2. What is the importance of data transfer in CAD/CAM systems?
3. Explain the data translation process.
4. Write the full form of the following:
   a. IGES
   b. STEP
   c. DMIS
   d. OOP
5. What are the different types of translators? Write down the difference between the two.
6. Compute the number of direct translators and neutral translators for five systems.
7. Explain the difference between the preprocessor and the postprocessor.
8. Discuss IGES file format. Write down its various benefits.
9. Briefly discuss the different sections of the IGES file.
10. Define STEP and discuss the benefits associated with the implementation of TEP.
11. Briefly explain the structure of the STEP file.

12. What is the difference between the head and data section of the STEP format?
13. What is the difference between the IGES and STEP file formats?
14. What do you mean by application protocols?
15. Which data specification language is used to represent the product information in STEP?
16. What do you mean by DMIS?
17. What are various benefits of the implementation of DMIS?
18. Write down the different components of the DMIS file.
19. Explain the concept of OOP.
20. Define the following:
    a. Objects
    b. Classes
    c. Data abstraction
    d. Inheritance
    e. Data encapsulation
    f. Polymorphism
21. Discuss the importance of OOP concepts in CAD/CAM/CMM integration.

References

1. S. Ranđelović and S. Živanović, CAD-CAM data transfer as a part of product life cycle, *Mechanical Engineering*, 2007;5(1):87–96.
2. H. K. Suhas, *From Quality to Virtual Corporation: An Integrated Approach*, Boca Raton, FL: CRC Press, 2000.
3. R. J. Goult and P. A. Sherar, *Improving the Performance of Neutral File Data Transfers*, Berlin: Springer, 1990.
4. E. A. Nasr and Ali K. Kamrani, *Computer-Based Design and Manufacturing: An Information-Based Approach*, Berlin: Springer, 2007.
5. B. William, Initial Graphics Exchange Specification IGES 5.3, ANS US PRO/IPO-100, 1996.
6. D. R. Lide, Initial Graphics Exchange (IGES) by Sharon J. Kemmerer, *A Century of Excellence in Measurements, Standards, and Technology*, Boca Raton, FL: CRC Press, pp. 246–249, 2001.

7. P. N. Rao, *CAD/CAM: Principles & Applications*, New Delhi: Tata McGraw-Hill Education (CAD/CAM systems), 784pp, 2004.
8. A. Papavasileiou, K. Gavros, V. Vasileiadis, and S. Savvidis, CAD/CAM—Interfaces: A Review. *2002 First International IEEE Symposium "Intelligent Systems" Proceedings*, vol. 3, pp. 378–382, September 2002.
9. S. M. Amaitik, 2005, Developing of a STEP feature-based intelligent process planning system for prismatic parts, Doctoral Dissertation, Middle East Technical University.
10. D. Loffredo, Fundamentals of STEP Implementation, STEP Tools, Inc., 12pp, http://www.steptools.com/library/fundimpl.pdf (accessed on June 8, 2015).
11. S. Jkemmerer, STEP: The grand experience, NISTSP939 report, July 1999.
12. R. J. Fischer, Parametric geometry creation methodology and utility for the STARS CFD Analysis Package, *ProQuest*, 2007; 102pp.
13. Steptools, http://www.steptools.com/support/stdev_docs/stpfiles/ap203/1797609in.stp (accessed on June 9, 2015).
14. K. Mills, Demystifying DMIS, http://www.qualitydigest.com/oct99/html/dmis.html (accessed on June 3, 2015).
15. SO 22093:2011, Industrial Automation Systems and Integration—Physical Device Control—Dimensional Measuring Interface Standard (DMIS), 2011, 708pp, http://www.mmsonline.com/articles/cad-based-measuring-software-handles-large-complex-parts (accessed on June 4, 2015).
16. Y.-J. Lin and P. Murugappan, A new algorithm for CAD-directed CMM dimensional inspection, *International Journal of Advanced Manufacturing Technology* 2000;16:107–112.
17. G. Hermann, *Advanced Techniques in the Programming of Coordinate Measuring Machines*, 978-1-4244-2106-0/08, IEEE, 2008.
18. S. V. B. Sathi and P. V. M. Rao, STEP to DMIS: Automated generation of inspection plans from CAD data, *5th Annual IEEE Conference on Automation Science and Engineering*, Bangalore, India, August 22–25, 2009.
19. B. Yuewei, W. Shuangyu, L. Kai, and W. Xiaogang, *A Strategy to Automatically Planning Measuring Path with CMM Offline*, 978-1-4244-7739-5/10, IEEE, 2010.
20. B. Squier et al., Dimensional Metrology Standards Consortium (DMSC, Inc.), CMM Quarterly Special Edition, 2014.

21. S. Horsfall, 2007, Step by Step DMIS Programming, CMM Technology Solution.
22. T. Budd, *An Introduction to Object-Oriented Programming*, 3rd Edition, Boston, MA: Addison Wesley Longman, 2002, http://www.cpp-home.com/archives/206.html (accessed on June 9, 2015).
23. A. Kamthane, *Object-Oriented Programming with ANSI and Turbo C++*, New Delhi: Pearson Education India, 2003, 728pp.
24. D. J. Eck and A. Pillay, eds. *Object-Oriented Programming*, Berlin: Springer, 2007.
25. L. Cardelli and P. Wegner, On understanding types, data abstraction, and polymorphism, *Journal of ACM Computing Surveys (CSUR)*—The MIT Press Scientific Computation Series, 1985;17(4):471–523.
26. M. P. Bhandarkar and R. Nagi, STEP-based feature extraction from STEP geometry for agile manufacturing, *Computers in Industry* 2000;41:3–24.

Chapter 5

Coordinate Measuring Machine

5.1 Introduction

Coordinate measuring machine (CMM) can be defined as a mechanical system that is mounted with the measuring probe to determine the point coordinates on the work piece surface. The advantage of CMM is that it enables the 3D measurement of the complex objects in minimum setups. In fact, the CMM plays a vital role in the mechanization of the inspection process [1], which is relatively a recent development in the measurement technology. It can be described as a very precise Cartesian robot equipped with a measuring probe, and used as a 3D digitizer [2]. The CMM produces a stream of x, y, z coordinates through contact or noncontact probes. The coordinate stream is interpreted by algorithms that support applications such as reverse engineering, quality control, and process control. The objective in the quality and process control is to determine whether the given object meets the design specifications, which is known as dimensional inspection. The CMMs are very versatile in their capability to record measurement of complex profiles with high sensitivity (0.25 μm) and speed.

The CMM comprises four major components: (1) the main structure consisting of three axes of motion, (2) probing system, (3) control or computing system, and (4) the measuring software. There have been wide ranges of designs and sizes for the CMM including the state-of-the-art probe technologies in the market. The most common applications of the CMM include dimensional measurement, geometrical tolerance measurement, and depth mapping. The measurement range of CMMs can vary from objects as small as circuit boards to parts as large as an engine. The CMM can be used in both metrology lab as well as the manufacturing floor. It can perform measurement in both point-to-point and scanning modes, depending on the complexity of the part geometries and the application requirement. In fact, the CMM establishes (X, Y, Z) axis coordinate of a point on the part surface of a part. The CMM accuracy has improved over the years, now reaching up to ±2 micron level. The dimensional inspection tasks carried out on the CMM can have measurement volumes up to 100 m^3. The advantages of the CMMs over the conventional measurement devices are flexibility, reduced setup time, improved accuracy, and improved productivity [3]. The CMMs find applications in many industries including aerospace, automotive, electronics, health care, plastics, and semiconductor. The CMMs are most suitable in the following scenarios:

- Multiple features are needed to be inspected (geometrical and dimensional).
- High flexibility is required due to the variety of components and multiple features.
- Reduce the unit cost by minimizing the rework or scrap cost.
- Production interruptions, that is, to inspect the first component before ramp up the production.

There are two types of coordinate systems in the world of measurement (or the CMM): (1) the machine coordinate system and (2) the part coordinate system. In the machine coordinate system, the X-, Y-, and Z-axes define the machine's motions. In the part coordinate, the three axes relate to the data or the features of the workpiece.

5.2 Main Structure

The structure of the CMM consists of three axes, each provided with a guide way to enable precise movement along the axes [4]. The guide ways are provided with a carrier that also moves along with the guide ways. The axes are fitted with a precision scale that records the position of the carrier with respect to the reference point. When the measuring probe makes contact with the measuring object, the measurement system records the position of all the three axes. There are many physical configurations of the CMMs depending on the method of moving the probe along the three axes relative to the object (Figure 5.1).

1. Cantilever type
2. Bridge type
3. Horizontal arm type
4. Column type
5. Gantry type

5.2.1 Cantilever Type

The probes are attached to the vertical quill that moves along the depth of the arm to achieve the Z-axis movement. The quill can also be moved along the length of the arm to achieve Y-axis, and the arm can be moved relative to the worktable to achieve the movement along X-axis. Broadly speaking, the measuring probe is attached to the Z-axis and moves in the vertical direction. The Z-axis carrier is fitted to the cantilever arm and provides movement in the Y-direction [4]. The Z-axis

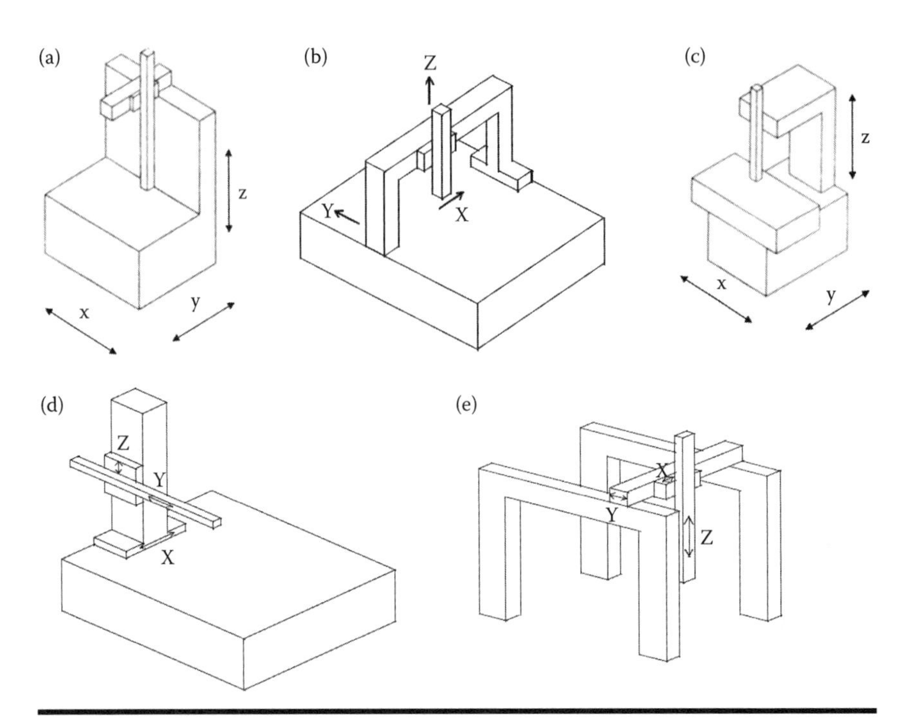

Figure 5.1 The different configurations of the CMM: (a) cantilever type; (b) bridge type; (c) column type; (d) horizontal arm type; and (e) gantry type. (Adapted from R. J. Hocken and P. H. Pereira, *Coordinate Measuring Machines and Systems, Second Edition,* Boca Raton, FL: CRC Press, 2016 ([accessed on July 8, 2015].)

movement is provided by the table. The parts larger than machine table can be inserted into the open side without restricting the complete machine travel. This type of CMM design provides easy access to the work area and has high workspace volume.

5.2.2 Bridge Type

In this type of CMM configuration, the arm is supported on both ends like a bridge and the table carries the quill (Z-axis), which can move in X-direction as well as Y-direction. This bridge construction can also be called as the traveling bridge. This type of bridge construction is very common in the industries. The measuring probe is attached to the Z-axis arm and the Z-axis carrier is attached to the X-axis arm. The X-axis carrier in turn is attached to the Y-axis arm. This type of design provides better rigidity and therefore the CMM has higher accuracy [4]. The limitation of this design is that it is difficult to place onto the worktable because of the obstruction from the vertical parts of the Y-axis frame.

5.2.3 Column Type

This kind of construction is very similar to the machine tool rather than the CMM. In fact, this design of the column type CMM is similar to the drilling machine or the vertical milling machine. The column type CMM achieves X- and Y-axes motion through the movement of the worktable. To achieve the Z-axis movement, the quill moves vertically. This type of CMM is commonly known as the universal measuring machine [4]. The construction of the column type CMM provides very good rigidity and high accuracy.

5.2.4 Horizontal Arm Type

The horizontal arm type CMM consists of a horizontal arm, which moves in the horizontal direction. The measuring probe is attached to the Y-axis arm and the main advantage of this type of CMM is that the work volume is large and free from any obstruction [4]. This type of machine can be used for large work pieces such as the car body.

5.2.5 Gantry Type

The gantry construction as shown in Figure 5.1e can be used for inspecting the large objects. The X- and Y-axes motion is achieved through the construction similar to the gantry crane. It means that the X- and Y-axes are placed overhead and supported by four columns from the base [4]. The probe quill can be moved relative to the horizontal arm to achieve the movement in

Z-direction. The main advantage of this type of configuration is that the operator can move with the probe, and measurements of large objects, such as the body of the car, can be carried out.

5.3 Probing Systems in Coordinate Measurement Machines

The probe can be defined as the sensing element that makes contact with the workpiece and records the readings [3]. It is one of the fundamental components of the CMM configuration. There are basically two types of probes, including the contact probe and the noncontact probe. The contact probes can further be categorized as the touch-trigger and the analog scanning type. The probe head is mounted at the end of one of the CMM's moving axes while the actual probe is attached to the probe head. The most common stylus tip is a ruby ball because it provides high wear life. The probing process can be divided into the following two tasks [5]:

1. Positioning: It is the process of bringing the work piece into measuring range of the probing system or vice versa.
2. Measuring: It is the process of comparing the dimensions of the actual part with that of the design specifications.

The environment where the probing system operates defines specific properties, which must be taken into account in order to get precise and accurate results. The requirements for the probing systems can be categorized into the application-specific properties, metrological characteristics, and system-specific properties as shown in Figure 5.2.

The application-specific requirements depend on the nature of the measurement task. For example, the scanning mode is a preferred choice for the sculptured surface as compared to the touch-trigger probe. Moreover, the contact probe finds it difficult, when it comes to measure the soft and fragile objects.

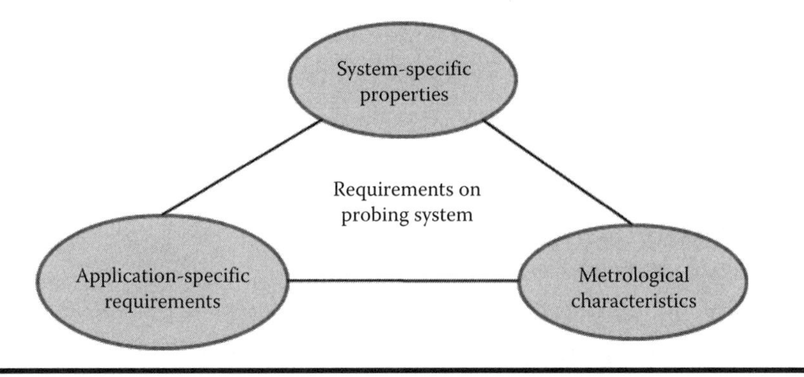

Figure 5.2 Requirements on the probing system. (Adapted from A. Weckenmann et al., *CIRP Annals—Manufacturing Technology,* 2004;53(2):657–684.)

The contact probe is limited by the size of the geometries and it is mainly used to inspect the regular shapes. The laser probe can also be used for the measurement of the free form surfaces. Since there is no contact between the laser scanner and the object surface, the problems of damaging the soft objects, measuring the small details, and capturing the complex free form surfaces can be eliminated. The disadvantages of the laser line scanners are limited accuracy and strong influence of surface quality. The measurement of the variety of features in different orientations requires a variety of different styli and accessory as well as articulating heads.

The metrological characteristics define the results that can be measured by a particular system. The reproducibility is the most important metrological property of the probing system. This is due to the fact that it is the only characteristic that cannot be improved by the qualification or compensation processes [5]. The reproducibility, sensitivity, backlash, pretravel, and overtravel defines some of the metrological characteristics of the probing system.

The system-specific properties include the economic aspects such as the costs, time, and additional equipment required for performing the measurements with the given probe. The weight of the probing system determines its dynamic properties and the static load on the measuring system. The size of the probing system has to be small enough so as not to obstruct the access to the workpiece features.

5.4 Application

With the growing use of GD&T, the CMM have become a significant contributor in the dimensional measurements. The CMM can be equipped with a variety of sensors and software to perform a wide range of measurement and inspection tasks, including the workpiece and layout inspection, reverse engineering, tool setup, process control, and archiving.

- *Measuring capacity*: CMM has been the ideal measurement system for small-to-medium sized components within the measuring range of up to X = 700 mm, Y = 700 mm, and Z = 700 mm.
- *Point verification*: To meet the demands of the current manufacturing technology, which requires accurate dimensional verification, CMM performs simple-to-complex point verifications.
- *Scanning*: With CMM, the complex surface profiling of components such as gas turbine airfoils can easily be performed.
- *Dimensional tolerances* (*DT*): The accurate and precise dimensional checks can be done effectively with the CMM unit.
- *Geometric tolerances* (*GT*): CMM can help in assessing the products critical contact areas and verify part curvature by scanning sample part profiles or surfaces.
- *Reverse engineering*: With up-to-date CMM technology, the sample parts can be scanned to obtain point cloud data, surfaces, or solid models to recreate lost prints or recreate parts for duplication while minimizing the engineering time and cost.

5.5 Virtual CMM

The concept of virtual coordinate measuring machine (VCMM) emerged in 1990s in order to predict the volumetric error of an inspection. The VCMM concept generally refers to the computer programs that simulate the behavior of the CMM and provide useful assistance in CMM-related tasks [6]. The VCMMs can be categorized into two groups. The first one includes the tools to analyze errors and uncertainties associated with the CMM inspection. The second group focuses on the simulation and representation of CMM inspection in cyber space and allows applications like operation training, intelligent inspection planning, offline programming, remote monitoring and controlling, among others. The VCMM typically creates a virtual replica of a physical CMM by modeling certain aspects of the characteristics of the CMM. The experiments on physical CMMs can be time-consuming and expensive, sometimes even risky when manually controlled by new operator, whereas in VCMMs all operations and CMM behaviors are virtualized in computer and therefore, the same task can be performed faster and risk-free. The reliability of the results generated by the VCMM can be validated by the experiments on the physical machine and can be improved through better model and simulation methods. The important applications of VCMM are as follows [6]:

- Error simulation and prediction, including the CMM geometry error, probing error, etc.
- Uncertainty evaluation of CMM measurement
- Online analysis of measurement uncertainty for CMM
- Offline accessibility analysis
- Offline programming of the CMM inspection
- Optimization of the CMM inspection plan
- Remote monitoring of the CMM inspection
- Remote control of CMM inspection

The VCMM consists of two basic modules including a module that simulates the residual and random errors depending on the kinematic system of the CMM and a module that simulates the work of the probe head [7]. The VCMM approach can be used to estimate the uncertainty for a particular measurement task on CMM using Monte Carlo simulation. Basically, virtual CMM (Figure 5.3) performs point by point simulation of measurements, emulating a measurement strategy and the physical behavior of CMM with dominating uncertainty contributions disturbing the measurement.

The simulated measurement should have all the facets of the real measurement that can significantly contribute to the measurement uncertainty. The important factors that affect the performance of the VCMM includes the systematic deviations of the slide ways, uncertainties due to calibration, thermal deformation of the slide ways, thermal expansion of the scales,

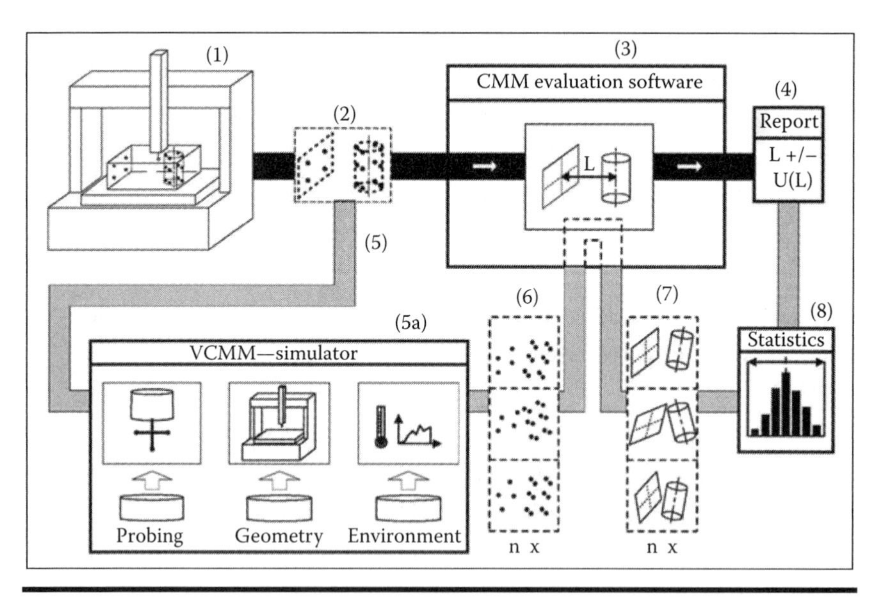

Figure 5.3 The virtual CMM concept. (Adapted from M. Trenk, M. Franke, and H. Schwenke, The "Virtual CMM": A software tool for uncertainty evaluation—Practical application in an accredited calibration lab; http://www.aspe.net/publications/Summer_2004/04SU%20 Extended%20Abstracts/Trenk-1670.PDF [accessed on July 8, 2015].)

drift effects, uncertainties of the stylus calibration, uncertainties when using a probing system with several styli, thermal expansion of the workpiece, and roughness of the work piece [8].

5.6 Application of the CMM in Statistical Quality Control

The CMM quality measurements depend on the environment and the process's stabilities for producing the accurate output from the captured data points. The CMM output should be accurate to give confidence to the user in order to either accept or reject the design. Therefore, CMM has to be used effectively and efficiently to get consistent and reliable results. However, the performance of CMM has limitations, which include the environmental conditions (cleanliness of the environment, temperature changes, etc.), background vibration levels, accuracy in measurement equipment (part sensor and encoders), and structural deflections between part sensor and encoders [9]. The other limitations include the type of CMM degree of freedom (DOF) and operating conditions, since it can introduce errors at the indicated position of the part sensor.

CMM can have statistical modules that can be used for statistical quality control. The purpose of statistical quality control is to check whether the manufacturing object meets its design specifications or not. The quality control is used to decide whether to accept the part design or not for use either individually or in an assembly. It can also be used to adjust machining parameters of manufacturing processes. The utilization of CMM enables the measurement of surface flatness, cylinder diameter, and cylindricity deviation and feature position with respect to the reference coordinate system. The integration of the measuring process with the quality and process control provides interface between the metrological functions of the measuring system with graphic representation. The charts including the R control chart (the gap), S chart (standard deviation), X chart

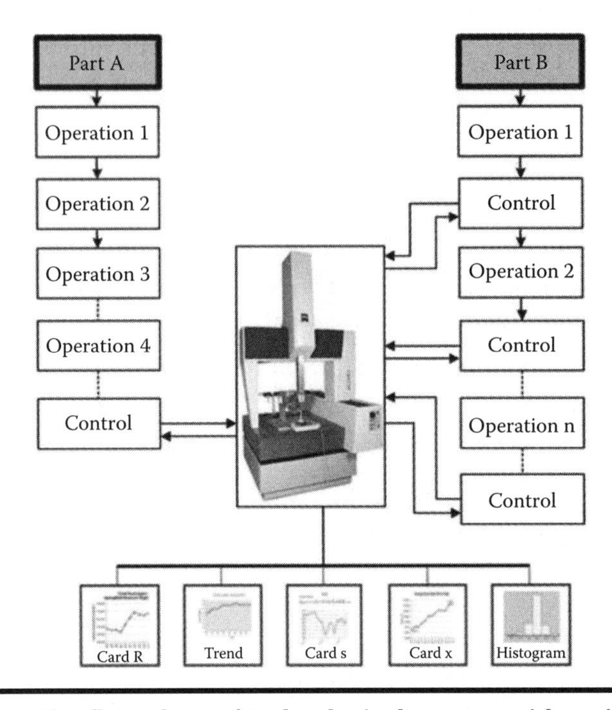

Figure 5.4 The flow chart of technological process with various control charts by using CMM. (Adapted from T. Kędzierski, A. Matusiak, and Poznan, *Application of Coordinate Measuring Machine in Statistical Control of Technological Process*, 1st International Workshop "Advanced Method and Trends in Production Engineering," University of Zilinia, Slovakia, 2004.)

(average value) as well as the trend and histogram can be generated using the data obtained from the CMM. The computational procedures are executed automatically and the results can be represented either in textual or graphical form [10]. The flow chart of the technological process with various control charts by using CMM is shown in Figure 5.4.

5.7 DMIS File Component

DMIS is a standard format of high-level programming language, which transfers the equipment and product definitions along with the process and recording the information necessary to perform G&DT measurements that employ coordinate metrology. DMIS contains definitions of the nominal features of the product, dimensional and geometric tolerances, feature constructions, part coordinate systems, and functional data. It communicates equipment definitions for various measurement resources, machine parameters, and measurement sensors. DMIS instructs touch probing and sensor's motions for manufacturing process validation and control [11].

DMIS is used for bidirectional transfer of inspection data between CAD systems and CMM. The generated program of measuring feature is executed to compare GD&T like position, orientation, size, and flatness, depending on allowable tolerances associated with feature [12].

5.7.1 Base Alignment

Datum alignment or part setup, as it is sometimes called, is required in order to set the position and orientation of CMM coordinate system. The part setup is usually defined by geometric tolerance definitions contained on component drawings.

5.7.2 Sensor Procedure

1. Sensor definition
   a. SNSDEF used for sensor definition.
      S(name) = SNSDEF/PROBE, type, CART, x, y, z, i, j, k, length, diameter
   b. Sensor definition using polar coordinate.
      S(name) = SNSDEF/PROBE, type, POL, rot1(A), rot2(B), i, j, k length, diameter
   c. Sensor definition using vectors.
      S(name) = SNSDEF/PROBE, type, VEC, i1, j1, k1, i2, j2, k2, length, diameter
   d. Sensor definition using vector direction.
      S(name) = SNSDEF/PROBE, S(name) or SA(name), VEC, i, j, k, length, diameter

e. Sensor definition using vector direction.
This syntax format uses a direction vector from the previously referenced sensor.

2. Calibration sensor
f. CALIB/SENS, SA(name), FA(name), n
CALIB is a command used for calibrating the previously defined sensor.
S(name) is the name of the sensor used to calibrate.
FA(name) is the name of a previously defined or measured feature that represents the calibration master artifact.

3. Selecting previously defined sensors
g. SNSLCT/SA(name)
Using SNSLCT statement is used to select a sensor with the SA option sensor names.

5.7.3 Feature Definition

F(name) = FEAT/type, Cartesian, direction

FEAT defined the name, coordinate information, position, and the normal vector direction of the feature. The coordinate information may be a Cartesian (rectangular) three-dimensional position (x, y, z), or a polar (circular) two dimensional position [12].

1. *Arc definition* using start and finish angles
F(name) = FEAT/ARC, dir, coordinate, x, y, z, i, j, k, radius, a1, a2
2. Circle definition
F(name) = FEAT/CIRCLE, dir, coordinate, x, y, z, i, j, k, diameter
3. Slot definition
F(name) = FEAT/CPARLN, dir, type, coordinate x, y, z, i1, j1, k1, i2, j2, k2 length, width
4. Cone definition
F(name) = FEAT/CONE, dir, coordinate , x, y, z, i, j, k, angle

5. Cylinder definition
   F(name) = FEAT/CYLINDER, dir, coordinate , x, y, z, i, j, k, diameter (,length)
6. Curve definition
   F(name) = FEAT/GCURVE, coordinate , x, y, z, i, j, k
7. Surface definition
   F(name) = FEAT/GSURF
8. Line definition using vectors
   F(name) = FEAT/LINE, UNBND, coordinate , x, y, z, i1 ,j1, k1, i2, j2, k2
9. Line definition using points
   F(name) = FEAT/LINE, BND, coordinate , x1, y1, z1, x2, y2, z2, i2, j2, k2
10. Plane definition
    F(name) = FEAT/PLANE, coordinate , x, y, z, i, j, k
11. Point definition
    F(name) = FEAT/POINT, coordinate , x, y, z, i, j, k
12. Sphere definition
    F(name) = FEAT/SPHARE, dir, coordinate , x, y, z (,i, j, k)

5.7.4 Feature Measuring

MEAS/Feat _type, F(name), n

MEAS statement is used to measure the previously defined feature with MEAS-ENDMES measurement block.

Feat _type: define the feature type.

F(name): define name of this feature in the previous definition, and n number of touch point.

5.7.5 Machine Movement

GOTO is a statement used to move CMM automatically from one measuring feature to other measuring feature in a way that avoids probe collision [12].

1. Cartesian
   GOTO/ x, y, z
2. Polar
   GOTO/POL, radius, angle, height
3. Incremental
   GOTO/INCR, distance, i, j, k
4. Defining a Cartesian home position
   FORM/x, y, z
5. Defining a polar home position
   FORM/POL, radius, angle, height
6. Moving to the home position
   GOHOME

5.7.6 Test of Geometrical and Dimensional Tolerance

- Geometrical dimensional: It contains upper limit and lower limit of dimension and acceptance depends on tolerance-value.
  T(name) = TOL/size_type, tolerance-value, direction
  Size_type: the dimension type required to measure like, length, width and height of the feature.
- Form tolerance: It is the tolerance of shape verification depending on datum face selected.
  a. Position tolerance:
     T(name) = TOL/POS, tolerance-value
  b. Perpendicularity tolerance:
     T(name) = TOL/PERP, tolerance-value, FA(datum)
  c. Parallelism tolerance:
     T(name) = TOL/PARLEL, tolerance-value, FA(datum)
  d. Angularity tolerance:
     T(name) = TOL/ANGLR, tolerance-value, FA(datum)
  e. Concentricity tolerance:
     T(name) = TOL/CONCEN, tolerance-value, FA(datum)
  f. Flatness tolerance:
     T(name) = TOL/FLAT, tolerance-value

g. Straightness tolerance:
   T(name) = TOL/ STRGHT, tolerance-value

5.7.7 Result Output

It describes how to associate tolerances with previously measured or constructed features in order to create reports.

- OUTPUT/FA(name), TA(name)
- FA(name): represents the name of the actual feature needed to be reported.
- TA(name): represents the predefined result tolerance used to report the feature(s).

5.8 Summary

The most popular inspection equipment, CMM has been discussed elaborately in this chapter. The different components of the CMM and its various configurations have been described. The role of the probing systems and the various applications of the CMM have been presented. The virtual CMM and the application of the CMM in the statistical quality control have also been discussed. The different components that make up the DMIS file have also been shown in this chapter.

QUESTIONS

1. Define coordinate measuring machine. Explain its working.
2. What are the different components of the CMM? Write down its various benefits over the conventional measuring devices.
3. When are CMMs most useful in the manufacturing environment?

4. What is the difference between the machine coordinate system and the part coordinate system?
5. Discuss the various configurations of the CMM.
6. Define the measuring probe. What is the difference between the contact and the noncontact probes?
7. What do you mean by the probing process?
8. Discuss the following characteristics of the probing system:
   a. Application-specific properties
   b. Metrological characteristics
   c. System-specific properties
9. Discuss the different applications of the CMM.
10. Explain the concept of the virtual CMM.
11. What are the different applications of the VCMM?
12. Discuss the role of the CMM in statistical quality control.
13. What do you mean by the DMIS? Why is it an important requirement for the CMM in the manufacturing industries?
14. Write down the syntax for the following DMIS components:
    a. Sensor definition for Cartesian and polar coordinates
    b. Circle and plane definition
    c. Tolerance definition
    d. Flatness and perpendicular tolerance
    e. Output definition

References

1. J. A. Bosch, Coordinate Measuring Machines and Systems, Boca Raton, FL: CRC Press, 1995, 496pp.
2. S. N. Spitz, Dimensional inspection planning for coordinate measuring machines, PhD thesis, University of Southern California, August 1999.
3. R. J. Hocken and P. H. Pereira, *Coordinate Measuring Machines and Systems, Second Edition*, Boca Raton, FL: CRC Press, 2016.
4. M. M. Ratnam, Coordinate Metrology Basic Dimensional Metrology, 2009, http://mechanical.eng.usm.my/MMR/MMR2/Chapter_7_-_Coordinate_metrology.pdf (accessed on July 8, 2015).

5. A. Weckenmann, T. Estler, G. Peggs, and D. McMurtry, Probing systems in dimensional metrology, *CIRP Annals—Manufacturing Technology*, 2004;53(2):657–684.
6. Y. Hu, Investigation and development of an advanced virtual coordinate measuring machine, PhD thesis, School of Engineering and Design Brunel University, May 2010.
7. J. Sładek, A. Gąska, M. Olszewska, R. Kupiec, and M. Krawczyk, Virtual coordinate measuring machine built using laser tracer system and spherical standard, *Metrology and Measurement Systems*, 2013;XX(1):77–86.
8. M. Trenk, M. Franke, and H. Schwenke, *The "Virtual CMM": A Software Tool for Uncertainty Evaluation—Practical Application in an Accredited Calibration Lab*; *Proceedings of ASPE: Uncertainty Analysis in Measurement and Design,* July 2004.
9. W. E. Singhose, W. P. Seering, and N. C. Singer, *The Effect of Input Shaping on Coordinate Measuring Machine Repeatability*, IFToMM World Congress on the Theory of Machines and Mechanisms, 1995, Milan, Italy.
10. T. Kędzierski, A. Matusiak, and Poznan, Application of coordinate measuring machine in statistical control of technological process, *1St International Workshop "Advanced Method and Trends in Production Engineering,"* University of Zilinia, Slovakia, 2004.
11. B. Squier et al., Dimensional Metrology Standards Consortium (DMSC, Inc.), CMM Quarterly Special Edition, 2014.
12. S. Horsfall, *Step by Step DMIS Programming*, CMM Technology Solution, 2007.

Chapter 6

Computer-Aided Inspection Planning

6.1 Introduction

The automation of the inspection process requires the development of a computer-aided inspection planning (CAIP). This is due to the fact that the CAIP acts as the integration link between the CAD/CAM and computer-aided inspection (CAI) [1]. The integration process links the CAD/CAM packages and the coordinate measuring machine (CMM) to enable communication between the two. The feature data of the part model such as basic features (hole, bosses, rib, etc.) as well as complex features (free-form surfaces) have to be extracted to develop the CAIP. This extracted data, which is used to generate the inspection path for the inspection planning, is then sent to the CMM. The extracted information is, in fact, utilized to make decisions regarding setup, probe selection and orientation, probe accessibility, and probing points, which ultimately leads to the development of CAIP (Figure 6.1). Moreover, the inspection plan (CAIP) is made up of two primary components: the global inspection planning (GIP) and the local inspection planning (LIP) [1].

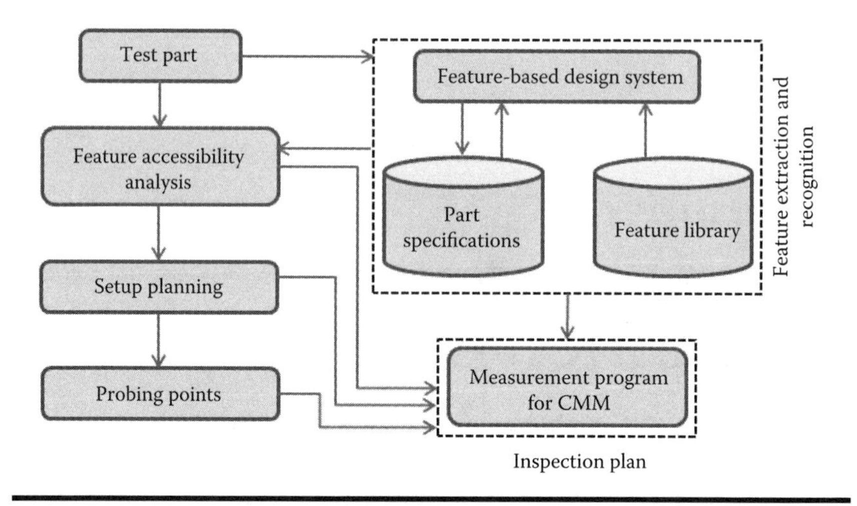

Figure 6.1 Manufacturing and geometric feature classification for inspection. (Adapted from G.-S. Yoon et al., *International Journal of Precision Engineering and Manufacturing,* 2004;5(1):60–68.)

The GIP includes the decisions such as the sequence of the setups and form features for the part, while the LIP comprises the determination and sequencing of measuring points on the surface. It can be concluded here that a well-established CAIP plays a significant role in enhancing the product quality and the manufacturing efficiency in the manufacturing process lifecycle.

6.2 Feature Extraction

The working of the feature extraction algorithms largely depends on the type of the data provided by the geometric modeler of the CAD system. The CAD modeler determines the method by which the user can create a design drawing and the data structure for storing the CAD model in the computer memory. Moreover, the data structure is determined by the types of information to be stored in the CAD database. Meanwhile, a number of systems and techniques can be used for the feature extraction depending on the application requirements. For example, the method based on an incremental feature converter utilizes both the feature information and geometric information for extracting the machining features from the design features [3]. These kinds of systems are based on an integrated geometric modeling system where both feature-based modeling and feature recognition can be carried out. These systems can handle feature interactions and protrusion features effectively by combining the capabilities of both the feature-based design and feature recognition. Similarly, the modular modeling approach [4] combines the feature recognition and feature-based design to integrate the design and the process planning processes. The system based on this approach is made up of four basic components: module creation, modular feature recognition, design with modules, and process planning preparation. This approach can overcome the problems associated with the unrecognized features and the inconvenience of using the individual basic machining features appeared due to the constraints of the traditional features recognition approaches and the traditional feature-based design approaches, respectively. However, this approach requires the part to be divided into several sections depending on the functional and geometric considerations (each of the parts is represented by the functional module). The functional module is selected from a module library for each section to complete the design of the entire part. The basic machining features are recognized in each module using the modular feature recognition method. The functional modules simplified the feature recognition because of the relatively limited and localized number of different features in a module. A method known as 3-space [5] can be used to obtain the 3D models of prismatic machine parts using 2D technical drawings. This method is based on the definition of the three surfaces base, top, and sweep base for determining the objects. The four different characteristics of the processing objects, namely external access direction, exit status, boundary geometry, and the processing order of the features has to be determined in this method. Initially, the object is represented as a prismatic raw material; then the features to be processed on it are removed; finally, the remaining object itself is created, thus generating the 3D view of the object.

A feature recognition system based on the intelligent feature recognition methodology (IFRM) has the ability to communicate with various CAD/CAM systems [6]. The methodology works particularly for the 3D prismatic parts created using a solid modeling package. The system based on this methodology consists of an input in the form of a neutral file in the initial graphics exchange specification (IGES) format. The information in this file has to be converted to the manufacturing information. The boundary (B-rep) geometrical information on the part design is analyzed by the feature recognition program and then the geometrical information based on a geometric reasoning approach is extracted using the object-oriented design software. The IFRM provides various benefits as follows:

- It provides a generic representation of both the simple and compound product, including the associated feature, geometry, topology, and the manufacturing data.
- It can also extract simple and curved features such as round corners.
- It has the flexibility to use different IGES file formats from various vendors offering different CAD systems.
- It is easily adaptable to many standard formats such as STEP or DXF.

The scale-space feature extraction technique [7] for feature extraction is based on the recursive decomposition of polyhedral surfaces into surface patches. These surface patches can be parameterized based on the local surface structure of the model to generate decompositions that correspond to the manufacturing, assembly or surface features relevant to the mechanical design. This technique automatically segments the 3D models (polyhedral representation) into features to be used for indexing, classification, and matching. This approach can be applied to the extraction of features that are invariant to the global structure of the model as well as small perturbations introduced due to the 3D laser scanning process using a new distance function defined on triangles instead of the points. There are also techniques that not only recognize manufacturing features, but also replicate, compound, and translate features [8]. The feature recognition processor based on this technique works by translating the design feature model of the part into the manufacturing feature tree using the properties of the design features and feature interacting relationships. In fact, the alternative interpretations of the manufacturing feature model for the parts are generated through various combinations, decomposition, tool approach direction operations, and the combination that provides the lowest cost is selected in the manufacturing tree. The hybrid (graph + rule-based) approach [9] for identifying the interacting features from B-rep CAD models of prismatic machined parts can handle both the noninteracting and interacting machining features. The hybrid approach addresses the variable topology features and handles both the adjacent and the volumetric feature interactions in order to provide a single interpretation for the volumetric feature interactions. A new concept of base explicit feature graphs and no-base explicit feature graphs can be used to depict the features having a planar base face-like pockets, blind slots, etc. and those without a planar base faces like passages, 3D features, conical bottom features, etc.

6.3 Computer-Aided Inspection Planning

The inspection planning is an integral part of the manufacturing and process planning because it involves some vital information that is critical in maintaining the quality of the manufacturing part. The essential steps of the inspection planning includes the selection of the most stable part orientation, identification of the number and distribution of the inspection points, feature accessibility analysis, sequencing of the probe orientations, removal of the duplicate faces, and sequencing of the faces. [10]. Broadly speaking, there are two stages, including the global inspection planning and the local inspection planning stages, that make up the overall inspection plan [11]. The global inspection planning stage consists of the information regarding the optimum inspection sequence of the features while the local inspection planning stage provides information related to the suitable number of measuring points and measuring point locations. The inspection plan can also be classified as the high-level inspection plan (HLIP) and the low-level inspection plan (LLIP) [12]. The HLIP consists of the determination of the setup of the part on the CMM table, probe to be used, and the orientation of the probe and the measurements to be performed with each setup, probe, and probe orientation. The LLIP consists of the sequence of operations with the high-level operations (change setup, change probe, etc.) and the low-level operations (move-CMM-arm). The techniques such as the clustering algorithm, path generation, and inspection process simulation [13] can be used for the tolerance feature analysis and accessibility analysis. This ultimately leads to the development of the inspection process plan directly from the CAD model. The Chavatal's greedy heuristic can also be utilized to minimize the cycle time for part setups and probe changes and the probe travel time for the required inspection features when the features are selected based on the tolerance specifications [14]. The traveling salesman problem (TSP) algorithm can also be applied to minimize the probe moving distance [15]. There have been many techniques that can be used to develop an automated inspection plan for the CMM. For example, the automated inspection plan (based on the CAD model path planning system) for the turbine blade surfaces on a three-axis CMM [16] achieves the precise and accurate inspection in automatic and time-efficient way. The CAD model of the workpiece and tolerance information can be used as an input to develop an algorithm for defining the points' accessibility [17]. This will automatically establish the accessibility domain of measurement points and group them into a set of clusters. The CAD model format is first changed to the stereo lithography (STL) or virtual reality model language (VRML) format for the probe orientation module (POM). The type of probes and measuring features are selected by the user and the coordinate systems are defined, tolerances, and data for measurement points are specified, and the probe paths are generated. The knowledge-based clustering algorithm [18] for the inspection plan works by grouping the inspection features into feature families and arranging the probe orientations into probe cells so that every inspection feature can be inspected at the assigned cell. The knowledge-based clustering algorithm incorporates the multiple constraints (such as determination of the feature waiting list and maximum probe number) and satisfies the requirement of grouping the inspection features. The technique provides better efficiency and effectiveness in probe selection and inspection planning for a CMM.

A functional tolerance model [19], which is based on the technological and topological-related surfaces (TTRS) and technologic product specification (TPS) methodologies, provides a complete framework for the inspection planning. This framework establishes a connection between the CAD and CAI. The functional tolerance model provides two types of information. The first one includes the definition of the fixture design activity and the second one is the definition of the information needed to identify the geometry to be inspected. The cascaded multisensor system significantly extended the capabilities of automated dimensional inspections of micro- and nano-structured components [20]. In the design-based inspection planning, the knowledge-based information is applied during the design stage. They also provide a novel and consistent overall view of the dimensional inspections of micro- and nano-structured components. The inspection workstation (IWS) used for investigating the knowledge requirements for the intelligent controllers [21] also presents a very good example of an automated inspection planning. The knowledge in this system refers to the information, data, processes, and other capabilities that are encoded or used by the controller and enables it to achieve its goals. A solid modeling server for intermediate geometric calculations is used to generate the inspection plan at the CMM from the manufacturing features of the part. The knowledge-based systems can effectively gather and use sensor information to overcome the volume and complexity of the required information for adequate performance [22]. The working of the knowledge-based inspection planning system for the solid objects is based on the following steps:

- Identification of the entities (edges, etc.) of the object to be measured
- Determination of the camera locations and viewing directions for performing the inspection
- Search for the entities and their inspection method once the sensor data is collected

6.4 Integration of Systems

The various geometric features are classified in the CAIP for the generation of the inspection steps. Although the inspected object is considered as the single feature in the CAD/CAM systems, for the measurement purpose, the inspection feature is classified into the surface information. The features in the inspection object can be divided into free-form geometries and analytic geometries as shown in Figure 6.2. The analytic geometries can further be

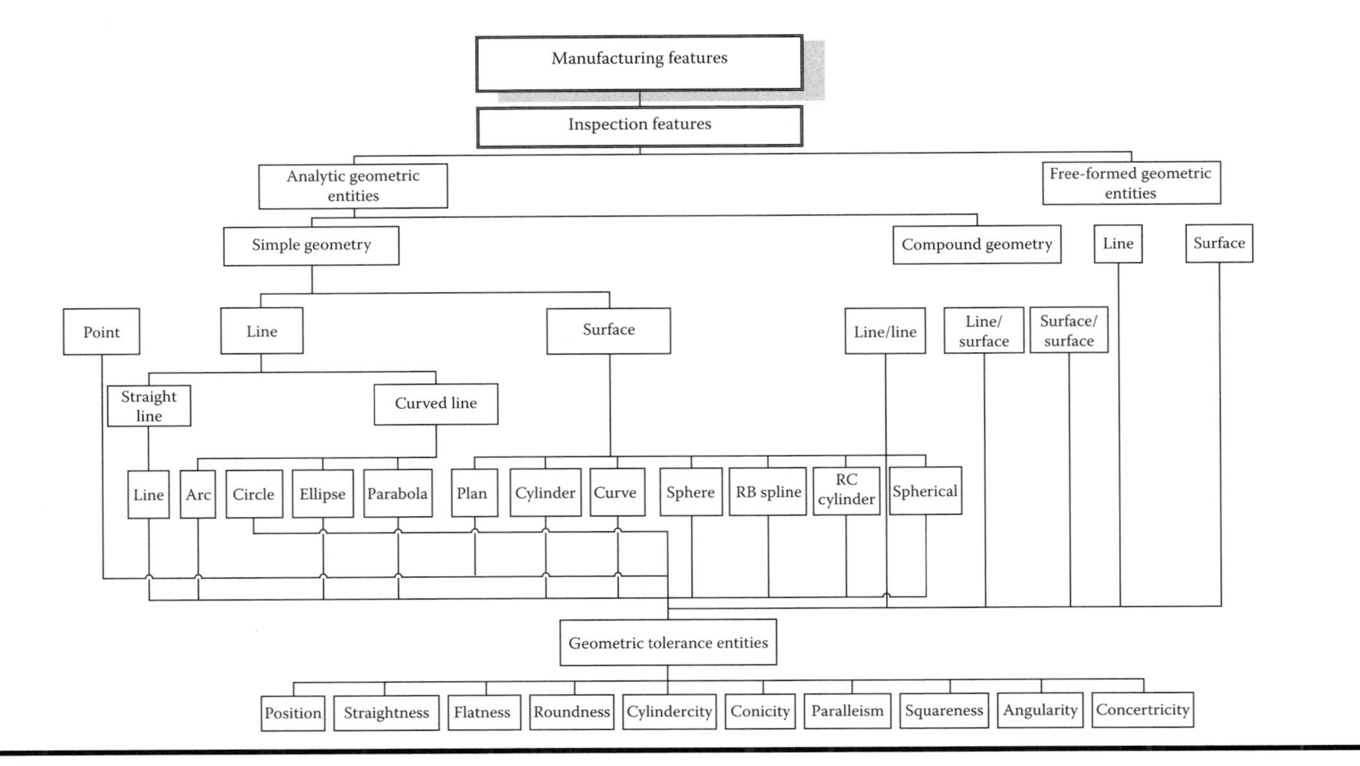

Figure 6.2 Feature classification for the development of the computer-aided inspection plan. (Adapted from E. A. Nasr et al., Set up planning for automatic generation of inspection plan, *The International Conference on Sustainable Intelligent Manufacturing*, Lisbon, Portugal, Jun 26–29, 2013.)

categorized into single geometry and integrated geometry. The single geometry consists of three entities: (a) point, (b) line that is classified into a straight line and a curved line, and (c) surface that is classified into the plane, cylinder, cone, and sphere. The inspection of a single surface requires six geometric tolerances, including the position, straightness, flatness, roundness, conicity, and cylidricity. However, for the inspection of the two or more single geometries (integrated geometries), four geometric tolerances, including the parallelism, squareness, angularity, and concentricity are required [23].

The inspection plan obtained from the CAIP module can be executed using the CMM. The inspection plan can be generated as a high-level inspection plan (HLIP) and a low-level inspection (LLIP) [12]. The HLIP determines how to set up the part on the CMM table, which probes to use and how to orient the probes, and which measurements to perform with each setup. Finally, the HLIP is expanded into a complete program for driving the CMM to inspect the object. The LLIP is a sequence of operations that contain the high-level operations (change setup, change probe) and the low-level operations. The LLIP also includes a complete path plan for the CMM and a work-piece localization process after each setup is performed. Finally, from the LLIP, a complete path plan for the CMM is obtained. The integration module between two or more systems helps to decrease the time associated with various processes in each system and aids to automate the system processes.

The hybrid knowledge-based approach that integrates the CAD and CAIP into computer-aided design inspection planning (CADIP) system has two primary interfaces to provide the link [24]. The integrated system achieves the ability to accept the part definition data from the CAD system and the ability to relate the inspection attributes to the geometric entities of the part. The knowledge-based system helps to define and structure the knowledge involved in the design and inspection processes [25]. The integration of the CAD and CAI systems in a CAIP system for on-machine measuring (OMM) is based on two stages: global inspection planning and local inspection planning [1,11]. The global inspection planning stage generates an optimum inspection sequence of the features by analyzing the feature information. The local inspection planning stage identifies the most stable part orientation, number and distribution of inspection points, feature accessibility analysis, sequencing of probe orientations, removal of duplicate faces, and finally sequencing of faces. The information model of a product model and a process model consisting of the defined data objects successfully achieves the direct integration of the dimensional inspection process into the production cycle [26]. This integration makes an efficient link of the design activity, the planning activity, and the inspection execution. Moreover, the product-based reasoning strategies requires the implementation of a computer-aided inspection process planning (CAIPP) system based on the feature recognition approach for the non-CMM-inspection and the environment of the generic computer-aided process planning support system (GCAPPSS) [27]. Although the CAD-directed inspection path planning for CMM is applicable to any object whose boundary is composed of planner, cylindrical, and conical faces [28], still the principles for the types of surfaces and the techniques for generating the measuring points has to be determined to generate the probe inspection path. This path should detect the possible collisions between the probe and the measuring object by applying a probe collision avoidance strategy for the given inspection path.

The development of an object-oriented inspection planning (OOIP) methods for generating the automatic inspection process planning using the CAD models is very popular [29]. The generative inspection process planner and an inspection path planner are implemented for the integration of a STEP model with the CMM. The inspection probe path consists of two parts: a local path, which represents the probe trajectory for measuring the individual features, and a global path, which represents the connections between the features. Similarly, the integrated system proposed by Spyridi [30] generates a high-level plan consisting of a collection of setups of the part, features to be inspected in each setup, and the probe selection and orientations to be used for each feature for inspecting 3D parts with CMM. The inspection plan is executed by inserting a geometric model of the part to be inspected together with a set of surface features and associated tolerances. The output includes the execution of the high-level plan by the CMM. The successful integration of the CMM into the CAD/CAM environment for generating the inspection path using CAD database of the part to be inspected can be realized in the works of Lin and Murugappan [31]. The applied algorithm generates the inspection path using the solid representations of the CAD/CAM system and then extended it to feature-based inspection planning. The tolerance specifications are also incorporated and a DMIS output is created for online analysis. A systematic methodology for feature identification and measurement analysis of different machined parts [32] has to be effective, flexible, and reprogrammable. A sensor-based automated machine vision system [32] with the optimal operating environment (achieved through Taguchi method) and online feature measurement (using the pattern matching method) represents an excellent example of the systematic methodology. The various steps include the image acquisition and digitization, image processing, feature extraction, and pattern recognition using feature parameter method. The approach includes the dimensional verification and shape matching of two-dimensional objects and can be used for subpixel accuracy for dimensional verification with minimal computation time. The standardization of the complete inspection process to the various ISO standards can be achieved through a STEP-compliant inspection approach [33]. It comprehensively includes all the information regarding the inspection work plan, working steps, mechanism to close the loop, and feedback of inspection results to the component model design. The standardization is achieved across the measuring and inspection across the total CAx chain through the integration of STEP-NC (ISO14649) along with AP219. This type of approach is especially applicable to the inspection of discrete components and consists of both CNC machine tools and CMMs in order to close the loop between the manufacturing and quality control. This novel inspection procedure provides a significant potential through automation of the inspection process and the integration of the standards and specifications. There are integrated systems that work in the feature-based inspection planning for the hybrid contact/noncontact inspection [34]. The feature-based inspection planning contains two modules: inspection method selection and inspection operations sequence optimization/planner. The application of this kind of methodology is crucial in industries for increasing the quality of the manufactured goods and reducing the product development cost. The characteristics of the sensors such as the occlusion, dimensions, and tolerance specifications are analyzed and included in an inspection-specific feature taxonomy organized in the form of a 3D matrix. This is particularly important to select the most suitable sensor for each feature. This helps to minimize the time and effort in terms of changing sensors, its orientation, and the corresponding part orientation.

6.5 Inspection Plan and Coordinate Measuring Machine

The inspection planning, which is obtained from CAIP, is exported to the CMM by generating the DMIS file. The DMIS is a programming language used for programming the mechanical, optical, laser, and video measuring systems.

The DMIS files must be translated to ensure that they can be used by the respective measuring machines, which are not DMIS-compatible. The translator tailored to relevant measuring software is referred to as the postprocessor. For example, all the information needed for the CMM program is combined in a device-independent DMIS format and then converted to equipment-specific programs [35]. This offline programming system for CMM uses a CAD model as an input and the user can select the required surface elements for the inspection. The system automatically generates the optimal distribution of the measuring points for the surfaces to be inspected and the local probe path. A heuristic algorithm is also applied to calculate the optimal sequence for the inspection task. The acquired and available information is combined in DMIS format to convert it into equipment-specific programs in the offline programming system [36]. The DMIS program developed by Yuewei et al. [37] could save about 30% of the time as compared to the conventional programming method. The methodology consists of automatically planning the measuring program by combining the measuring points and the path control points. The system proposed by Sathi and Rao [38] used the STEP (ISO 10303) file of the part to be inspected as an input and provide the inspection process plan of the part as an output in DMIS (ISO 22903) format. It is a very good example of the automated generation of process plans for the part inspection from the CAD models using the CMM. The system has three basic steps:

1. Use of geometric and product information of the part in order to identify the surface features to be probed for inspection
2. Automation of the setup planning and probe selection accounting for the feature accessibility
3. The probe path planning and generation of the inspection plan

An efficient and collision-free path can be generated for the CMM to inspect a collection of points using a practical path planner comprising both the sequencing and collision avoidance [39]. The issues of the measurement error (due to imperfections of the measuring device and the imperfection with geometric characteristics of the measured feature) in automated inspection of the free-form surfaces defined by cubic B-spline can be handled by the analytical models [40]. The need of probe accessibility for the dimensional inspection plan using CMM [41] represents another scenario where inspection plan and the CMM are dependent on each other. The accessibility analysis of inner features such as slots and holes is very important to study the influences of the probe length and volume. Similarly, the feature-based technique to determine a desired number of measuring points and accuracy generate the inspection plan to be executed on CMM [42]. The integration of the inspection activity with design and planning helps the users to effectively control the devices from CMM computers to design and planning computers [43]. The accessibility problem of a slot and other similar features such as pockets and slot blind can be solved efficiently using the mathematical approaches [44]. Therefore, the mathematical methods based on the analytical equations can help in the integration of the inspection and the CMM. A system known as feature-based inspection and control system (FBICS) for machining and inspection of mechanical parts is proposed by the Kramer et al. [45] to carry out automated hierarchical process planning. In this system, the CMM used the feature-based description for the shape of the object as an input to the machining or inspection center. The system provides the ability to generate plans for many levels of the hierarchical control system and implemented the two-stage planning. The system serves the following purposes:

- It demonstrates the feature-based inspection and control.
- It serves as a test bed for solving various issues in feature-based manufacturing.
- It also tests the usability of STEP methods and models.

According to Limaiem and ElMaraghy [46], the automated inspection planning and offline programming requires the integration of at least of the following main tasks:

- Accessibility analysis of the part and measurement of points
- Clustering of measurement points
- Sequencing of measurement points
- Collision-free probe path generation

The strategy introduced by the Spyridi and Requicha [47] for the inspection planning is based on the local and global accessibility cones and algorithms for computing and clustering of direction cones. These algorithms are based on the computation of Gaussian images and Minkowski (or sweeping) operations. The direction of cone clusters is needed to generate the minimal set of directions for inspecting the part. The direction cone can be defined as a set of directions that characterize the accessibility of a feature on a workpiece using a CMM touch probe. The different accessibility analysis techniques that can be applied differs in power and computational cost and fails in certain circumstances. This means that the entire feature cannot be measured with a single straight probe and the user either has to backtrack, segment the feature, and analyze the accessibility of the subfeatures or implement a method capable of dealing with bent probes. Similarly, Spyridi and Requicha [48] develop a system to automatically generate a high-level program for inspecting parts with CMM. The system uses a solid model of a tolerance part as an input to get information related to the number of setups, probe selection, and orientations associated with the surface features to be inspected. The problem is represented in the form of a state where each state consists of a collection of the measurements and constraints on these measurements. The state contains a set of inspection plans and the state has to evolve iteratively with the selection and application of operators in order to eliminate invalid and expensive plans from the state. The program establishes the geometric and optimization constraints and produces a valid and nearly optimal plan. The planner combines AI techniques with powerful algorithms to precede the state of the art in CMM planning and offline programming in many directions:

- The strategy is flexible and is not limited to a fixed sequence of planning operators.
- The same constraints lead to more accurate and efficient plans.
- The features can be segmented and replaced in some cases by the CMM table or fixture surfaces, for example, in situations when they cannot be inspected entirely with a single probe orientation.
- The accessible directions can be computed efficiently and accurately.

6.6 Coordinate Measuring Machine

The four different criteria, including the least squares circle, minimum zone circle, maximum inscribed circle, and minimum circumscribing circle, can be used to study the effect of CMM point coordinate uncertainty [49]. Dhanish and Mathew [49] found that the random uncertainties associated with the CMM measurements increases if the circularity error derived from the measurement points and the error is found to be proportional to the measurement uncertainty. Raghunandan and Venkateswara [50] predicted the flatness error accurately at reduced sample sizes in batch and mass production set-ups. They presented the methodologies and strategies for the measurement of flatness error estimation using CMM. The determination of the flatness error is based on the computational geometry techniques applied to the minimum zone solution method, ensuring that the measurement is consistent with the guidelines laid down by the standards for flatness error inspection. They found that the flatness error estimation achieved at sufficiently reduced sample sizes has much better degree of accuracy (90% and above), thus leading to the sufficient reduction in inspection time and cost. A strategy for the evaluation of flatness error using form tolerances for the control of manufactured surfaces has also been introduced [51]. They found that the surface finish can also be included as a parameter along with other parameters such as tolerance band, part geometry, and surface area for determining the best sample size. They also found that it is better to obtain the required number of inspection points for flatness error measurement on the basis of surface quality instead of sample size for measurement in order to achieve better accuracy. A hybrid inspection planning can also be developed for CMM-based measurements combining a laser line scanner and a touch trigger probe [52]. The inspection features being measured are identified and constructed based on the part's CAD model. Subsequently, a knowledge-based sensor selection approach is applied to choose the most suitable sensor for each specified inspection feature. The two inspection planning modules—laser scanning module and tactile probing module—are developed to plan the inspection process automatically for all the specified inspection features. Finally, the measured data are collected and evaluated for tolerance verification. The hierarchical planning system uses heuristics for the path planning in dimensional inspection [53]. It includes the automatic selection and modification of probe angle based on a local accessibility analysis and simulated the collision-free inspection path in a CAD environment before it does measurement on the CMM. The planning system has following features:

- It is based on 3D planning system.
- It can detect collisions.

- It can deal with complex parts having sculptured surfaces and multiple features.

A proper methodology is crucial for the definition of the touch probe required to measure the different points during the automatic inspection of mechanical parts [54]. This is due to the fact that it allows the determination of the probe's characteristics such as the maximum length of the stylus and sufficient number of styli. It automatically acquires the number and orientation of the styli from the CAD model successfully and accurately. This method can also be used to identify the position and orientation of the part on the CMM table. To reduce the uncertainties associated with the touch probe, they should be qualified using a unique reference sphere like a single sphere. The effect of the probe size and measurement strategies also affect the measurement uncertainty of the CMM [55]. Therefore, it is very important to devise a proper inspection plan (including appropriate probe size and dimensions) keeping in mind the application requirement.

6.7 Literature Classifications

The classification table for the various strategies reviewed in the literature is shown in Table 6.1.

6.8 Summary

It can be concluded that different methodologies and techniques can be utilized for feature extraction and recognition, inspection planning, and integration of different systems. However, there exists very few works related to the extraction of GD&T using STEP as the input file format. The IGES file format has also been not considered

Table 6.1 Literature Review Classification

Authors	*Feature Extraction*	*Inspection Planning (IP)*	*IP by Integrating Two or More Systems*	*Inspection Planning Using CMM*	*Coordinate Measuring Machine*	*Technology Considered*
Lee and Kim [3]	/					Integrated geometric modeling
Tseng [4]	/					Conventional feature-based design approach and the conventional feature recognition approach
Cayiroglu [5]	/					2D technical drawings
Abouel Nasr and Kamrani [6]	/					Intelligent feature recognition methodology (IFRM)
Bespalov et al. [7]	/					Local surface structure
Li et al. [8]	/					The properties of the design features and feature interacting

(*Continued*)

Table 6.1 (*Continued*) Literature Review Classification

Authors	*Feature Extraction*	*Inspection Planning (IP)*	*IP by Integrating Two or More Systems*	*Inspection Planning Using CMM*	*Coordinate Measuring Machine*	*Technology Considered*
Sunil et al. [9]	/					Hybrid (graph and rule-based) technique
Beg and Shunmugam [10]		/				An object-oriented methodology
Cho et al. [57]		/				And local inspection planning stages
Zhang et al. [13]		/				The clustering algorithm, path generation and inspection process simulation
Hwang et al. [14]		/				Chavatal's greedy heuristic
Fu et al. [15]		/				Travelling salesman problem (TSP)

(*Continued*)

Table 6.1 (*Continued*) Literature Review Classification

Authors	*Feature Extraction*	*Inspection Planning (IP)*	*IP by Integrating Two or More Systems*	*Inspection Planning Using CMM*	*Coordinate Measuring Machine*	*Technology Considered*
Chang and Lin [16]		/		/		CAD model path planning system
Marefat and Kashyap [22]		/				Knowledge-based
Hunter et al. [19]		/				(TTRS) and (TPS)
Linss et al. [20]		/				Knowledge-based information
Messina et al. [21]		/				Knowledge-based
Vafaeesefat and ElMaraghy [17]		/		/		Knowledge-based
Ajmal and Zhang [18]		/				Knowledge-based
Steven [12]		/		/		(HLIP) and (LLIP)
Yoon et al. [23]			/			Knowledge-based

(*Continued*)

Table 6.1 (*Continued*) Literature Review Classification

Authors	*Feature Extraction*	*Inspection Planning (IP)*	*IP by Integrating Two or More Systems*	*Inspection Planning Using CMM*	*Coordinate Measuring Machine*	*Technology Considered*
Adil and Ketan [24]			/			A hybrid knowledge-based approach
Lee et al. [1]			/			Global inspection planning and local inspection planning
Cho et al. [11]			/			Global inspection planning and local inspection planning
Wong et al. [27]			/			The product-based reasoning strategies
Barreiro et al. [26]			/			Implementation of a knowledge-based system
Fan and Leu [58]			/			Applying a probe collision avoidance strategy

(*Continued*)

Table 6.1 (*Continued*) Literature Review Classification

Authors	*Feature Extraction*	*Inspection Planning (IP)*	*IP by Integrating Two or More Systems*	*Inspection Planning Using CMM*	*Coordinate Measuring Machine*	*Technology Considered*
Hunter et al. [25]			/	/		Knowledge-based
Lin and Murugappan [31]			/	/		Integration of the coordinate measuring machine (CMM) into the CAD/CAM environment
Gu and Chan [29]			/			An object-oriented inspection planning (OOIP)
Spyridi [30]			/	/		Knowledge-based
Bhattacharyya [32]				/		A systematic methodology
Ali et al. [33]				/		A STEP-compliant inspection approach

(*Continued*)

Table 6.1 (*Continued*) Literature Review Classification

Authors	*Feature Extraction*	*Inspection Planning (IP)*	*IP by Integrating Two or More Systems*	*Inspection Planning Using CMM*	*Coordinate Measuring Machine*	*Technology Considered*
Mohib et al. [34]				/		A feature-based inspection planning
Spitz and Requicha [39]				/		The multiple goals technique
Wu et al. [41]				/		Applying a probe collision avoidance strategy
Jiang and Chiu [42]				/		A feature-based technique
Barreiro et al. [43]					/	The STEP standard philosophy
Chiang and Chen [44]					/	The analytical equations
Kramer et al. [45]				/	/	Feature-based inspection

(*Continued*)

Table 6.1 (*Continued*) Literature Review Classification

Authors	*Feature Extraction*	*Inspection Planning (IP)*	*IP by Integrating Two or More Systems*	*Inspection Planning Using CMM*	*Coordinate Measuring Machine*	*Technology Considered*
Lee and Mou [40]				/		Cubic B-spline
Limaiem and ElMaraghy [46]				/		A heuristic algorithm
Hermann [35]				/		A heuristic algorithm
Spyridi and Requicha [47]				/		Gaussian image
Sathi and Rao [38]			/	/	/	A hybrid knowledge-based approach
Spyridi and Requicha [48]				/		Gaussian images and Minkowski (or sweeping) operations
Hermann [36]				/		A heuristic algorithm
Yuewei et al. [37]				/		Automatically inspection planning generation

(*Continued*)

Table 6.1 (*Continued*) Literature Review Classification

Authors	*Feature Extraction*	*Inspection Planning (IP)*	*IP by Integrating Two or More Systems*	*Inspection Planning Using CMM*	*Coordinate Measuring Machine*	*Technology Considered*
Dhanish and Mathew [49]					/	Monte Carlo simulation technique
Raghunandan and Rao [50]					/	The minimum zone method (MZM)
Zhao et al. [52]					/	A hybrid inspection planning
Yau and Menq [53]					/	Local accessibility analysis
Dépincé [54]					/	Automatic inspection
Keferstein et al. [56]		/			/	Accuracy of a multisensor CMM
Rajamohan et al. [55]					/	The effect of probe size and measurement strategies on the CMM

for the GD&T extraction in the previous work. A complete methodology for the automation of the inspection process is hardly found in the history. Finally, very few works concerning the detailed procedure for setup planning, integration of the CAD/CAIP systems, DMIS application have been identified.

QUESTIONS

1. Explain the development of the computer-aided inspection plan (CAIP) with a schematic diagram.
2. Describe the two basic components of CAIP.
3. Write down the different factors that influence the feature extraction algorithms.
4. Discuss the method based on incremental feature converter.
5. Discuss the 3-space method, which is used to obtain the 3D models of the prismatic parts.
6. Write down the characteristics and the various benefits of intelligent feature recognition methodology (IFRM).
7. Define scale–space feature extraction technique and discuss its applications.
8. What do you mean by the modular modeling approach for the feature extraction?
9. Discuss briefly the hybrid (graph + rule-based) approach for feature extraction.
10. Write down the important steps of the inspection planning.
11. Why are Chavatal's greedy heuristic and the traveling salesman problem required in the inspection planning?
12. Briefly discuss the working of the knowledge-based clustering algorithm.
13. Describe the characteristics of the functional tolerance model.
14. Write down the various steps of the knowledge-based inspection planning system for the solid objects.
15. How can the different features be classified in the CAIP?

16. Discuss the integration of the CMM into the CAD/CAM systems.
17. What do you mean by the feature-based inspection planning?
18. How the inspection plan can be executed on the CMM?
19. What do you mean by the offline programming system? What is the basic requirement for the offline programming? Write down its various benefits.
20. What do you mean by the accessibility analysis? Define direction cone.
21. How are analytical methods important to the inspection planning?
22. Discuss the hierarchical process planning.
23. How is surface finish important to the evaluation of the flatness error?
24. Write down the importance of measurement uncertainties to the development of the automated inspection plan.

References

1. H. Lee, M.-W. Cho, G.-S. Yoon, and J.-H. Choi, A computer-aided inspection planning system for on-machine measurement—part I: Global inspection planning. *KSME International Journal*, 2004;18:1349–1357.
2. E. A. Nasr, A. Al-Ahmari, O. Abdulhameed, and S. H. Mian, Set up planning for automatic generation of inspection plan, *The International Conference on Sustainable Intelligent Manufacturing*, Lisbon, Portugal, Jun 26–29, 2013.
3. J. Y. Lee and K. Kim, A feature-based approach to extracting machining features. *Computer-Aided Design*, 1998;30(13):1019–1035.
4. Y. J. Tseng, A modular modeling approach by integrating feature recognition and feature-based design, *Computers in Industry*, 1999;39:113–125.
5. I. Cayiroglu, A new method for machining feature extracting of objects using 2D technical drawings, *Computer-Aided Design*, 2009;41:1008–1019.

6. E. A. Nasr and A. K. Kamrani, A new methodology for extracting manufacturing features from CAD system, *International Journal of Computer and Industrial Engineering*, 2006;l(51):389–415.
7. D. Bespalov, W. C. Reglia, and A. Shokoufandeh, Local feature extraction and matching partial objects, *Computer-Aided Design*, 2006;38(P):1020–1037.
8. W. D. Li, S. K. Ong, and A. Y. C. Nee, Recognizing manufacturing features from a design by feature model, *Computer Aided Design*, 2002;34(P):849–868.
9. V. B. Sunil, R. Agarwal, and S. S. Pande, An approach to recognize interacting features from B-Rep CAD models of prismatic machined parts using a hybrid (graph and rule based) technique, *Computers in Industry*, 2010;61(P):686–701.
10. J. Beg and M. S. Shunmugam, An object oriented planner for inspection of prismatic parts—OOPIPP, *International Journal of Advanced Manufacturing Technology*, 2002;19:905–916.
11. M.-W. Cho, H. Lee, G.-S. Yoon, and J. Choi, A computer aided inspection planning system for on-machine measurement-part II: Local inspection planning, *KSME International Journal*, 2004;18(8):1349–1357.
12. N. S. Steven, Dimensional inspection planning for coordinate measuring machines, Doctoral dissertation, University of Southern California, 1999.
13. S. G. Zhang, A. Ajmal, J. Wootton, and A. Chisholm, A feature-based inspection process planning system for co-ordinate measuring machine (CMM), *Journal of Materials Processing Technology*, 2000;107(P):111–118.
14. C. Y. Hwang, C. Y. Tsai, and C. A. Chang, Efficient inspection planning for coordinate measuring machines, *International Journal of Manufacturing Technology*, 2004;23(P):732–742.
15. M. W. Fu, S. K. Ong, W. F. Lu, I. B. H. Lee, and A. Y. C. Nee, An approach to identify design and manufacturing features from a data exchanged part model, *Computer Aided Design*, 2003;l(35):979–993.
16. H.-C. Chang and A. C. Lin, Automatic inspection of turbine blades using a 3-axis CMM together with a 2-axis dividing head, *International Journal of Advanced Manufacturing Technology*, 2005;26:789–796.
17. A. Vafaeesefat and G. A. ElMaraghy, Automated accessibility analysis and measurement clustering for CMMs, *International Journal of Production Research*, 2000;38(10):2215–2231.

18. A. Ajmal and S. G. Zhang, The application of a knowledge based clustering algorithm as an aid to probe selection and inspection process planning, *Proceedings of Institution Mechanical Engineers: Journal of Engineering Manufacture*, 1998;212(B):299–305.
19. R. Hunter, M. Guzman, J. Moller, and J. Perez, A functional tolerance model: An approach to automate the inspection process, *Journal of Achievements in Materials and Manufacturing Engineering*, 2008;31(2):662–670.
20. G. Linss, S. C. N. Topfer, and U. Nehse, Automatic execution of inspection plans for knowledge-based dimensional measurements of micro and nanostructured components, *XVIII IMEKO World Congress Metrology for a Sustainable Development* September 17–22, Rio de Janeiro, Brazil, 2006.
21. E. Messina, J. Horst, T. Kramer, H. Huang, T. Tsai, and E. Amatucci, *A Knowledge-Based Inspection Workstation*, National Institute of Standards and Technology, 1993.
22. M. Marefat and R. L. Kashyap, *Planning for Inspection Based On CAD Models*, 0-8186-2910-X/92 $3.00 0, IEEE, 1992.
23. G.-S. Yoon, G.-H. Kim, M.-W. Cho, and T.-I. Seo, A study of on-machine measurement for PC-NC system, *International Journal of Precision Engineering and Manufacturing*, 2004;5(1):60–68.
24. M. B. Adil and H. S. Ketan, Integrating design and production planning with knowledge-based inspection planning system, *The Arabian Journal for Science and Engineering*, 2005;l(30):2B.
25. R. Hunter, M. Guzman, J. Möller, and J. Perez, Implementation of a tolerance model in a computer aided design and inspection system, *Journal of Achievements in Materials and Manufacturing Engineering*, 2006;17(1–2):345–348.
26. J. Barreiro, J. E. Labarga, A. Viza´n, and J. Rıos, Information model for the integration of inspection activity in a concurrent engineering framework, *International Journal of Machine Tools & Manufacture*, 2003;43:797–809.
27. F. S. Y. Wong, K. B. Chuah, and P. K. Venuvinod, Automated inspection process planning: Algorithmic inspection feature recognition, and inspection case representation for CBR, *Robotics and Computer-Integrated Manufacturing*, 2006;22:56–68.
28. K.-C. Fan and M. C. Leut, Intelligent planning of CAD-directed inspection for coordinate measuring machines, *Computer Manufacturing System*, 1998;II(l-2):43–51.

29. P. Gu and K. Chan, Generative inspection process and probe path planning for coordinate measuring machines, *Journal of Manufacturing Systems*, 1996;15(4):240–255.
30. A. J. Spyridi, Automatic generation of high-level inspection plans for coordinate measuring machines, Doctoral Dissertation, University of Southern California, 1994.
31. Y.-J. Lin and P. Murugappan, A new algorithm for CAD-directed CMM dimensional inspection, *International Journal of Advanced Manufacturing Technology*, 2000;16:107–112.
32. B. K. Bhattacharyya, *Dimensional Inspection and Feature Recognition through Machine Vision System*, 978-1-4244-6484-5/10/$26.00 ©, IEEE, 2010.
33. L. Ali, S. T. Newman, and J. Petzing, Development of a STEP-compliant inspection framework for discrete components, *Proceedings of the Institution of Mechanical Engineers, Part B: Journal of Engineering Manufacture*, 2010;219:557–563.
34. A. Mohib, A. Azab, and H. ElMaraghy, Feature-based hybrid inspection planning: A mathematical programming approach, *International Journal of Computer Integrated Manufacturing*, 2009;22(1):13–29.
35. G. Hermann, Advanced techniques in the programming of coordinate measuring machines, *Applied Machine Intelligence and Informatics*, 2008. SAMI 2008, IEEE, pp. 327–330.
36. G. Hermann, Feature-based off-line programming of coordinate measuring machines, *INES Proceedings of the IEEE International Conference on Intelligent Engineering Systems.* September 15–17, 1997, Budapest, Hungary, pp. 545–548.
37. B. Yuewei, W. Shuangyu, L. Kai, and W. Xiaogang, A strategy to automatically planning measuring path with CMM offline, *Proceedings of International Conference on Mechanic Automation and Control Engineering, (MACE)*, IEEE, 2010, Wuhan, China, pp. 3064–3067.
38. S. V. B. Sathi and P. V. M. Rao, STEP to DMIS: Automated generation of inspection plans from CAD data, *5th Annual IEEE Conference on Automation Science and Engineering*, Bangalore, India, August 22–25, 2009.
39. S. N. Spitz and A. A. G. Requicha, Multiple-goals path planning for coordinate measuring machines, *Proceedings of the 2000 IEEE International Conference on Robotics & Automation*, April 2000.

40. G. Lee and J. Mou, Measurement error reduction in automated inspection of free-form surfaces defined by cubic B-spline, *Journal of Manufacturing Processes*, 2000;(2)3:174–186.
41. Y. Wu, S. Liu, G. Zhang, Improvement of coordinate measuring machine probing accessibility, *Precision Engineering*, 2004;28:89–94.
42. B. C. Jiang and S.-D. Chiu, Form tolerance-based measurement points determination with CMM, *Journal of Intelligent Manufacturing*, 2000;13:101–108.
43. J. Barreiro, S. Martinez, J. E. Labarga, E. Cuesta, Validation of an information model for inspection with CMM, *International Journal of Machine Tools & Manufacture*, 2005;45:819–829.
44. Y. M. Chiang and F. L. Chen, CMM probing accessibility in a single slot, *International Journal of Advanced Manufacturing Technology*, 1999;15:261–267.
45. T. R. Kramer, H. Huang, E. Messina, F. M. Proctor, H. Scott, A feature-based inspection and machining system, *Computer-Aided Design*, 2001;33:653–669.
46. A. Limaiem and H. A. ElMaraghy, Automatic planning for coordinate measuring machines, *Proceedings of the 1997 IEEE, International Symposium on Assembly and Task Planning*, Marina del Rey, California, August 1997.
47. A. J. Spyridi and A. A. G. Requicha, Accessibility analysis for the automatic inspection of mechanical parts by coordinate measuring machines, *Proceedings of the IEEE International Conference on Robotics & Automation*, 1990, Cincinnati, OH, pp. 1284–1289.
48. A. J. Spyridi and A. A. G. Requicha, Automatic programming of coordinate measuring machines, *Proceedings of IEEE International Conference on Robotics and Automation*, 1994, San Diego, CA, pp. 1107–1112.
49. P. B. Dhanish and J. Mathew, Effect of CMM point coordinate uncertainty on uncertainties in determination of circular features, *Measurement*, 2006;39:522–531.
50. R. Raghunandan and P. Venkateswara Rao, Selection of an optimum sample size for flatness error estimation while using coordinate measuring machine, *International Journal of Machine Tools & Manufacture*, 2007;(47):477–482.

51. R. Raghunandan and P. V. Rao, Selection of sampling points for accurate evaluation of flatness error using coordinate measuring machine, *Journal of Materials Processing Technology*, 2008;202:240–245.
52. H. Zhao, J. Kruth, N. V. Gestel, B. Boeckmans, P. Bleys, Automated dimensional inspection planning using the combination of laser scanner and tactile probe, *Measurement*, 2012;45:1057–1066.
53. H.-T. Yau and C.-H. Menq, Path planning for automated dimensional inspection using coordinate measuring machines, *Proceedings of the 1991 IEEE, International Conference on Robotics and Automation*, Sacramento, CA, April 1991.
54. P. Dépincé, Visibility: An application to a probe's definition, *Proceedings of the Institute of Mechanical Engineers, Part B: Journal of Engineering Manufacture*, 1999;213(7):747–750.
55. G. Rajamohan, M. S. Shunmugam, G. L. Samuel, Effect of probe size and measurement strategies on assessment of freeform profile deviations using coordinate measuring machine, *Measurement*, 2011;44:832–841.
56. C. P. Keferstein, M. Michael, G. Reto, T. Rudolf, J. Thomas, A. Matthias, J. Becker, Universal high precision reference spheres for multisensor coordinate measuring machines, *CIRP Annals - Manufacturing Technology*, 2012;784–788.
57. M.-W. Cho, H. Lee, G.-S. Yoon, and J. Choi, A feature-based inspection planning system for coordinate measuring machines, *The International Journal of Advanced Manufacturing Technology*, 2005;26(9):1078–1087.
58. K.-C. Fan and M. C. Leu, Intelligent planning of CAD-directed inspection for coordinate measuring machines, *Computer Integrated Manufacturing Systems*, 1998;11(1–2):43–51.

Chapter 7

Automatic Feature Extraction

7.1 Introduction

Computer-aided process planning (CAPP) plays a very significant role in the integration of computer-aided design (CAD) and computer-aided manufacturing (CAM) systems. However, even with advances in manufacturing and automation, the seamless integration is still not possible due to improper flow of information between the CAD and CAM systems. Moreover, the CAD system includes geometric information of a part, which are not compatible for the downstream applications such as process planning and inspection. The different CAD or geometric modeling packages store the design information in their own propriety formats [1]. The structures of these databases are different from each other. According to Yakup Yildiz [2], an intelligent interface between CAD and CAPP systems is essential to carry out precise process planning. This is due to the fact that the working of CAPP system, to a very large extent, relies on the accurate data derived from CAD systems. Therefore, the feature recognition techniques provide such an intelligent link between CAD and

CAPP systems. The establishment of such a link is not easy because the functioning of CAD and CAPP systems is based on different databases. The CAD database is geometry-based, which consists of entities such as points, lines, and arc, while CAPP systems are based on features including faces, cylinders, grooves, and pockets. To overcome such an obstacle, the concept of automatic feature extraction (AFE) technique was introduced. There have been many techniques, including graph-based and hint-based, such as cell division, cavity volume, convex hull, and lamina slicing, that can be used in AFE.

7.2 Automatic Feature Extraction

Automatic feature extraction (AFE) has been identified as an important element of CAPP. It can be defined as a methodology that helps in the development of process plans automatically [3]. In fact, it provides a platform to accomplish a coordinated design and planning activities for the realization of the final product. The working of AFE system is based on the following basic steps [3] as shown in Figure 7.1:

- Data file translator
- Part form feature classifier
- Manufacturing operation selector

The first module translates a CAD data file into an object-oriented data structure. A file in initial graphics exchange specification (IGES) or the standard for the exchange of product model data (STEP) format is used as input to the AFE system. The second step involves the classification of different part geometric features using the object-oriented data. Therefore, in this module, the translation of the information in the file is carried out to classify all the features of the component. The final step involves the selection of the manufacturing operation to perform the manufacturing processes to

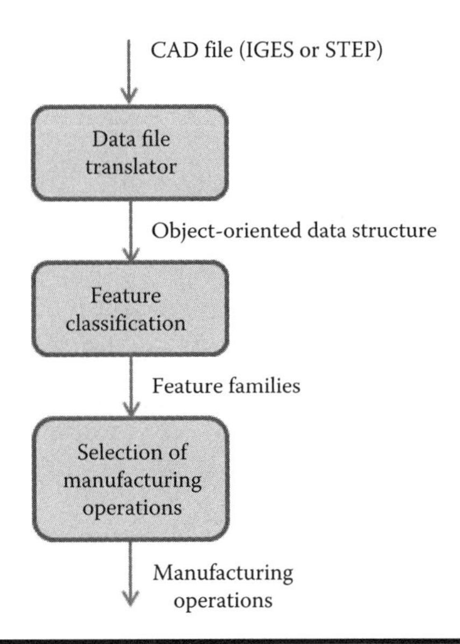

Figure 7.1 Basic AFE system. (Adapted from S.-C. Liu, M. Gonzalez, J.-G. Chen, *Computers in Industry,* 1996;29:137–150.)

machine the different features. There have been many algorithms that can be used for extracting machining features from the CAD file [4,5]. These algorithms greatly depend on the type of data provided by the geometric modeler of the CAD system. The different algorithms can be described as follows:

- Syntactic pattern recognition approach: This approach utilizes semantic primitives for the analysis of a component [6]. Moreover, in pattern recognition approach, primarily, the symmetric part designs are used for feature recognition [7]. The main components of the syntactic pattern recognition method are an input string, a pattern grammar, and a parser. In this approach, the features of the part design are represented by using some semantic (pattern) primitives and these semantic (pattern) primitives are written in a string form. A set of grammars and rules is used to define a particular character string. A parser is used to decide if the input character string is true or not.

- Logic-based approach: Logic-based approach is useful for the feature recognition of the B-rep and the CSG models [6]. In the B-rep model, the part is described in the form of low level entities such as faces, edges, and vertices [8]. A higher level is required for performing part recognition and this higher level is known as a feature. Each feature can be recognized by a separate rule. Features are recognized one by one in sequence until the whole set of features of the solid model is achieved. In the CSG representation model, the part design is represented in terms of a solid model. This solid model consists of solid primitives (block, cylinder, sphere, torus, wedge, and cone) where the solid primitive (feature) is considered as a higher level of representation [9,10].
- Graph-based approach: The topological relationships between geometric entities of a CAD model can be represented using a graph approach. The simple heuristics can be applied to the graph to identify the features. In this approach, the B-rep of a part is translated into a graph whose nodes and arcs represent the information of a part. The feature recognition is achieved by identifying a subgraph from the shape of a designed object where the subgraph is isomorphic to the feature graph.
- Section techniques: This method is typically used for tool path generation and automatic process planning for 2.5D components encountered in the aerospace and avionics industries. The technique involves validating the 2.5D components by testing the workpiece for approach direction and the presence of nonplanar faces. The part is checked for presence of sharp corners, fillets, chamfers, hidden surfaces, tolerance requirements, and cutter interferences with the fixture and with the component. The part volume is sliced in the X–Y plane to get the machining profiles and intersection curves.

- Volume decomposition and composition approach:
  In this approach, the geometry decomposition algorithms are applied on 3D volumes obtained through Boolean operations in the CSG model. The objectives of the volume decomposition are to decompose a volume to be machined, identify material to be removed from a base stock, and break down this volume into nonoverlapping units corresponding to distinct machining operations. There are two different approaches: convex hull decomposition and volume decomposition.
- Convex hull algorithm: The convex hull decomposition technique decomposes a volume by subtracting it from its convex hull and repeating the process for all the resulting volumes. The algorithm, however, has a problem of nonconvergence, resulting in erroneous volumes of features [11].
- Cell decomposition: These methods have been applied for the determination of machining volumes from stock and part models. The Boolean difference between the volume of the stock and the volume of the final part yields the total volume to be removed. The volume is then decomposed into individual pieces corresponding to specific machining operations.
- Expert system approach: An expert system (ES) is a computer program designed to carry out the problem solving capabilities of the human mind. It consists of two main elements:
  - Production knowledge
  - An inference engine

The knowledge and experience of an expert are converted to a set of facts and rules used to solve the problem. In the field of feature recognition, the expert system is used to determine how to manufacture a solid object. The production knowledge includes the procedural knowledge and the declarative knowledge [12]. The procedural knowledge involves the production rules related to the generation of the process plan from the part design. The production rules are usually expressed in the form of condition–action pairs, if this condition occurs, then do this action, and otherwise do that action [13].

7.2.1 Feature Extraction and Recognition

The different steps for features extraction and classifications can be described as follows [6]:

Step 1. Extract the geometry and topology entities of the designed object model from IGES file or STEP file format.
Step 2. Extract topology entities of each of the basic surface and identify its type.
Step 3. Test the feature's existence on the basic surface based on loops.
Step 4. Identify feature type.
Step 5. Identify the detailed features and extract the related feature geometry parameters.
Step 6. Extract all GD&T test faces depending on the functionality of the part.
Step 7. Identify the detailed machining information for each feature and the designed part.

7.2.1.1 IGES File Format

The part design in CAD software is represented as a solid model using the constructive solid geometry (CSG) technique. Moreover, the solid model of the part design consists of solid primitives and their combinations. The CAD software generates and provides the geometrical information of the part design in the form of an ASCII file (IGES), the standard format. The boundary representation (B-rep) geometrical information of the part design can be analyzed using a feature recognition program that is created specifically to extract the features from the geometrical information based on the geometric reasoning object-oriented approaches. Finally, a rapid prototype has to be produced to compare the features between the CAD model and the prototype model which verifies the quality of the rapid prototype.

The prismatic parts for feature extraction can be created using a Mechanical Desktop 6 power pack® CAD system or any other CAD software that supports IGES file format translators (B-REP Solid (186) with Analytical Surfaces). The feature recognition program has to be developed using Windows-based Microsoft Visual C++ 6 on a PC environment. The extracted entities can include vertices, edges, loops, and faces. Feature recognition involves the identification and grouping of feature entities from a geometric model [6]. The GD&T can also be extracted from CAD model and exported in an IGES file format. The IGES file format export the GD&T as a general note by displaying all datum, all geometrical tolerance test type and their values, the boundary of faces for every test and their data.

The boundary (B-rep) geometrical information of the part design has to be analyzed using a feature recognition program that is created specifically to extract the features from the geometrical information based on the geometric reasoning and object-oriented approaches. The feature recognition program can recognize these features: slots (through, blind, and round comers), pockets (through, blind, and round comers), inclined surfaces, holes (blind and through), and steps (through, blind, and round comers). These features are called manufacturing information and are mapped to process planning as an application for CAM. The structure of feature extraction methodology has been shown in Figure 7.2. The intelligent feature recognition methodology (IFRM) consists of three main phases: (1) a data file converter, (2) an object form feature

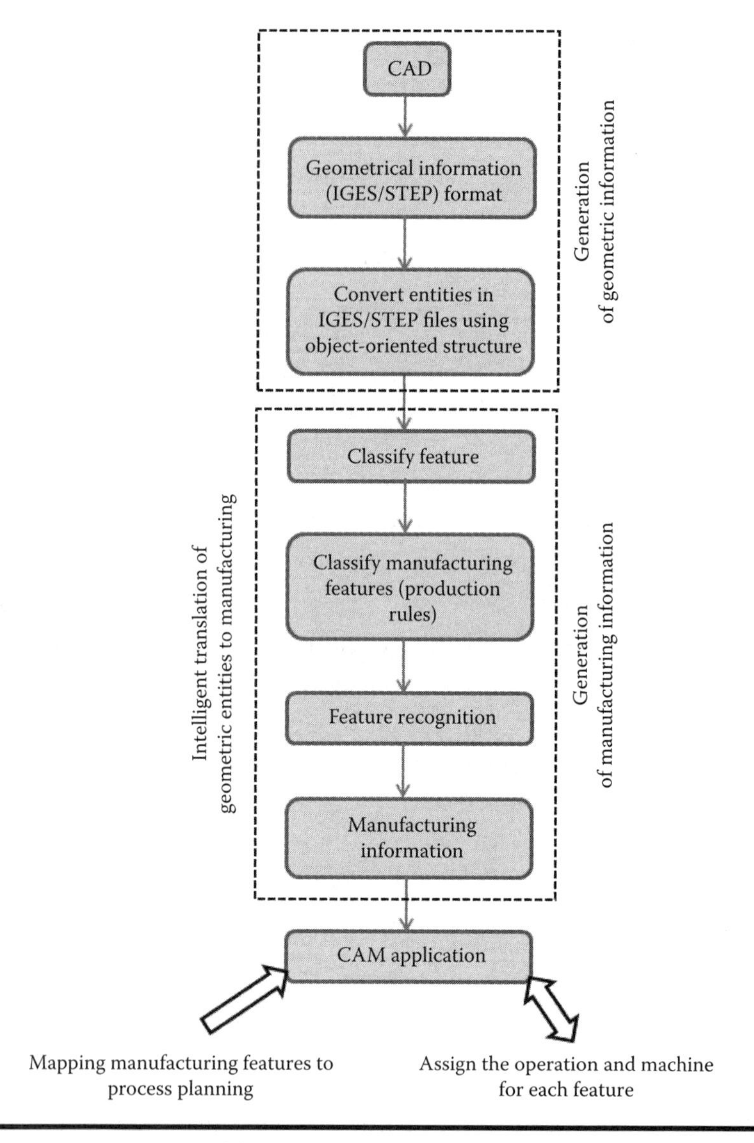

Figure 7.2 Structure of feature extraction methodology. (Adapted from E. A. Nasr and A. K. Kamrani, *Computer-Based Design and Manufacturing: An Information-Based Approach*, Berlin: Springer, 2007.)

classifier, and (3) a manufacturing features classifier (production rules). The first step toward automatic feature extraction is achieved by extracting the geometric and topological information from the (IGES/STEP) CAD file and redefining it as a new object-oriented data structure as demonstrated in Figure 7.3 [6].

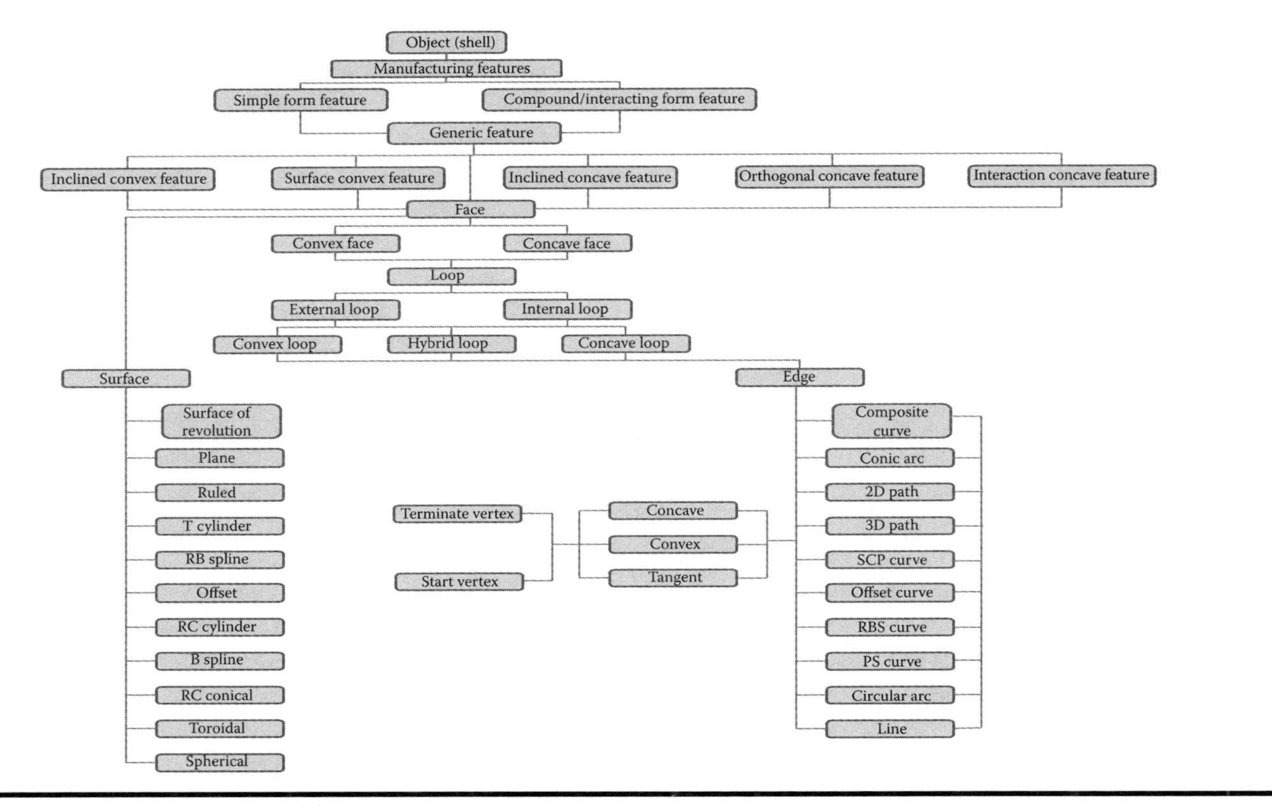

Figure 7.3 Hierarchy of classes and attributes of the designed object. (Adapted from E. A. Nasr and A. K. Kamrani, *Computer-Based Design and Manufacturing: An Information-Based Approach*, Berlin: Springer, 2007.)

7.2.1.2 STEP File Format

The feature extraction of the prismatic parts is achieved by exporting the CAD model created using CATIA V5 R21 in a STandard for the Exchange of Product data (STEP). The STEP file format is exported at AP203 that includes configuration controlled for the 3D designs of mechanical parts and assemblies (ISO10303-203:1994), which is one of the most widely used application protocols of STEP [14,15]. The feature recognition program can be developed using Windows-based Microsoft Visual C++ 6 on a PC environment. The AP-203 Edition 2 is a recently released new version of the AP-203 standard for exchanging 3D geometry between CAD systems and one of the extensions, which includes geometric dimensions and tolerances (GD&T) data [16]. The structure of STEP AP-203 can be understood from the chart shown in Figure 7.4.

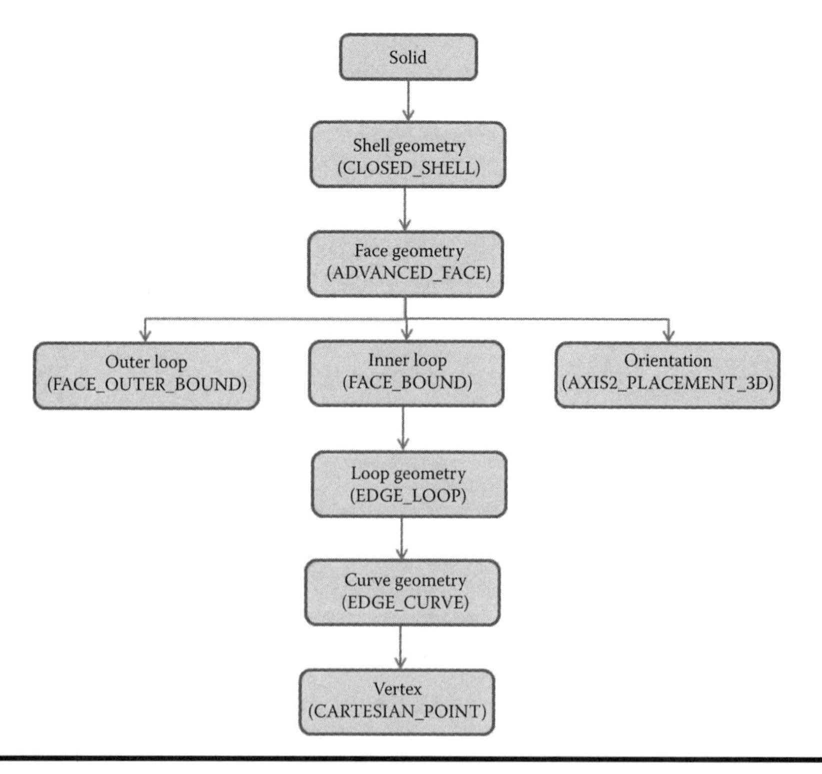

Figure 7.4 Structure of the STEP AP-203.

The STEP AP-203 represents 3D model in B-rep format. The solid B-rep entity represents the complete definition and topology of the geometry in the solid B-rep format. The outer extent of the solid is defined by closed shell. The closed shell consists of faces, which are defined by an advanced face. Each face is represented by both outer loop and inner loops, which, in turn, are defined with face outer bounds and face bounds, respectively. The face bounds are defined by edge loops and contain an edge loop that bound the face. An edge loop consists of oriented edges. The edges in turn consist of vertices, which are represented by vertex points. The 3D model STEP files are accessed to obtain the relevant geometric and topological information. The geometric data of the model is used for subsequent steps of the feature recognition.

7.2.2 Depression Features

7.2.2.1 Depression Features (Single)

Feature faces can be recognized successfully using the following algorithm for depression features [17]:

1. Find out the minimum and maximum x, y, and z value.
2. Extract the faces that have a constant x or y or z value throughout the edge loop but this value should be greater than the minimum value and less than the maximum value for these coordinates in case of plane faces (plane depression faces).
3. Extract those faces also in which all coordinate values are changing throughout the edge loop (inclined faces).
4. In case of plane and circular faces found in STEP file, extract not only the plane depressed faces according to step 2 but also the circular faces that have a common edge with the depressed faces (for round corner features).
5. The common edges between the extracted plane depressed faces are plane line edges, the common line

edges between plane and circular faces are tangent line edges, and the common line edges between circular and circular faces are also tangent line edges.

7.2.2.2 Depression Features (Multiple)

The multiple features present within the component can be recognized using the following steps:

6. Extract all the faces according to the previous algorithm.
7. Check the face connections and make face groups.
8. Apply recognition rules to all the groups and recognize each feature separately.

7.2.3 Feature Classification

The form feature can be classified in the multilevel form feature taxonomy based on the feature geometry and topological characteristics. The hierarchy of form feature has been described in Figure 7.5.

7.2.4 Feature Recognition Rules

For each feature shown in Figure 7.6, there is a specific production rule that defines how the feature can be extracted. For example, the following algorithm can be used for the extraction of the slot blind feature shown in Figure 7.7.

7.2.4.1 Slot Blind Feature

The recognition of the slot blind feature requires checking the following condition:

- If faces are connected with one another such that facel and face2 have three common edges of type line and face3 and face4 have two common edges of type line in their edge loops

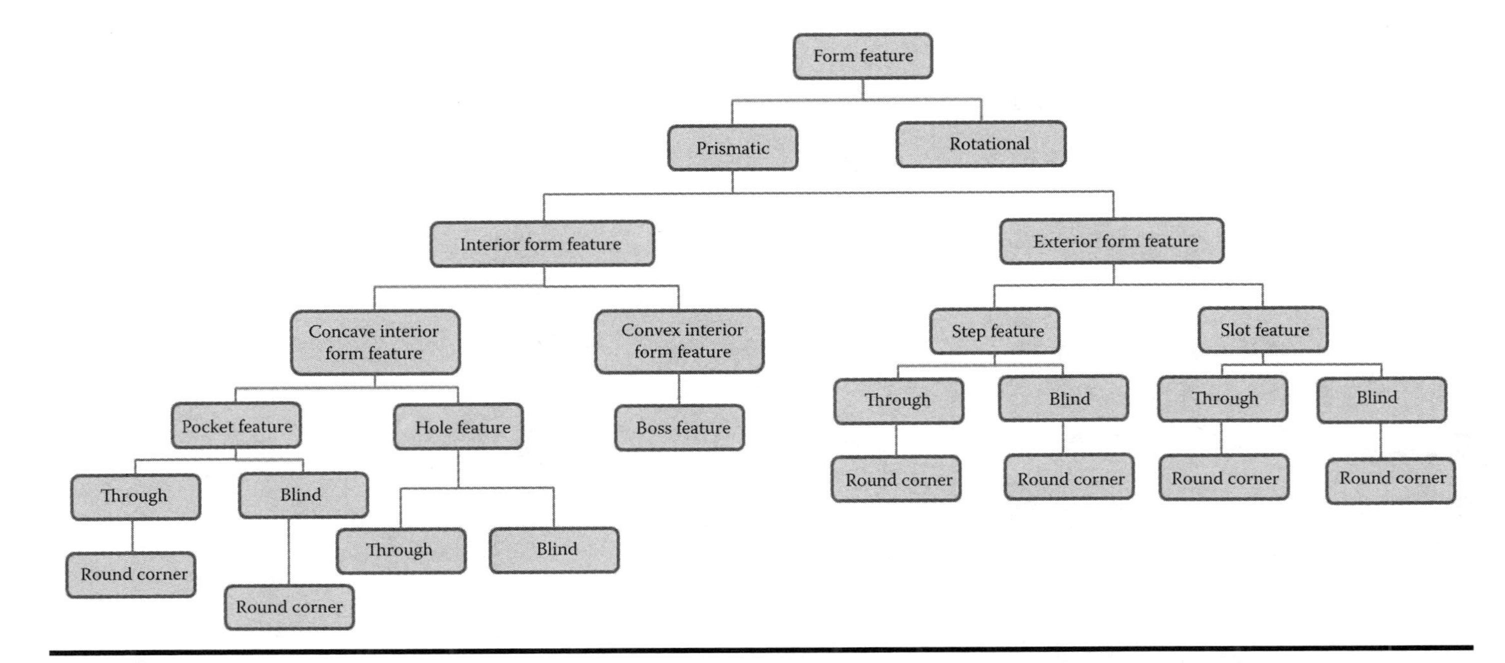

Figure 7.5 Hierarchy of form features. (Adapted from E. A. Nasr and A. K. Kamrani, *Computer-Based Design and Manufacturing: An Information-Based Approach*, Berlin: Springer, 2007.)

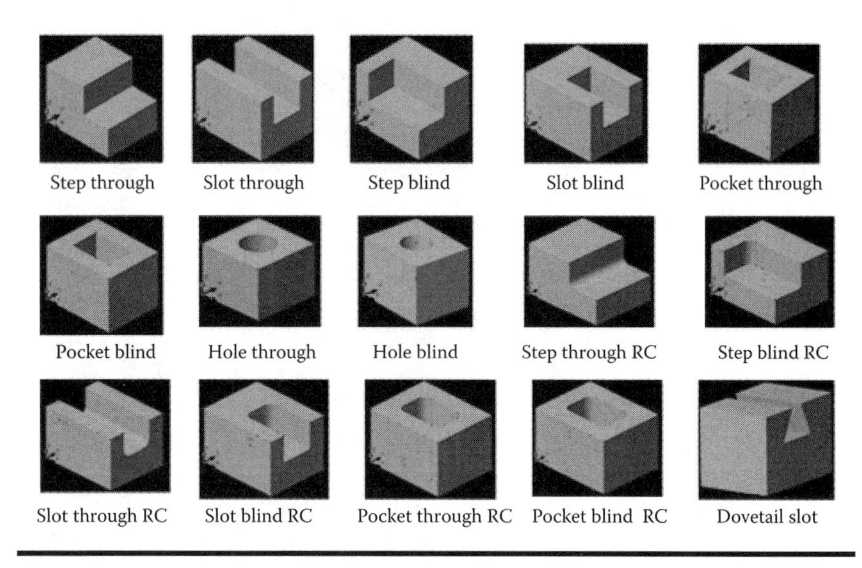

Figure 7.6 Manufacturing features. (Adapted from E. A. Nasr and A. K. Kamrani, *Computer-Based Design and Manufacturing: An Information-Based Approach*, Berlin: Springer, 2007.)

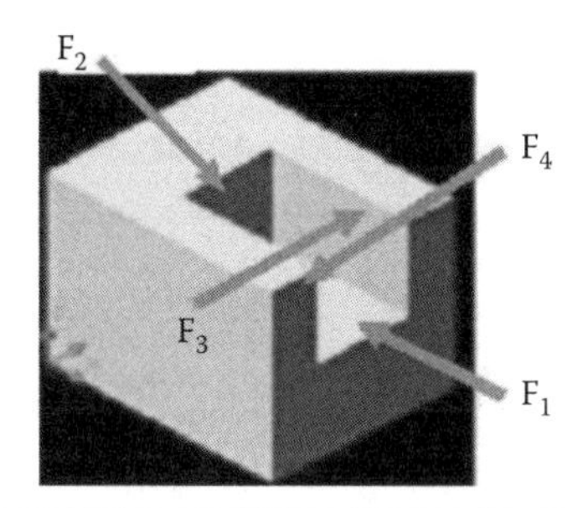

Figure 7.7 Slot blind.

- If face1 is perpendicular to face2, face3, and face4
- If face3 and face4 are parallel to each other

7.2.5 GD&T Extraction

7.2.5.1 GD&T Extraction in IGES File Format

The GD&T can also be extracted from the CAD model [18], which is exported in IGES translator and does not support tolerances but it is treated as a note in the exported IGES file by displaying all datum faces, all geometrical tolerance test types, its value, the boundary of faces for every test, and its datum.

Algorithm for Datum Extraction

1. Open IGES file
2. Create PSection file (store PEntry objects)
3. While Read line from PSection file
   3.1. If there are more than 4 line of 110 PEntery followed by 212 line
      3.1.1. If the last value contains 2H
         3.1.1.1 Take as datum name
         3.1.1.2 Take the last point and Go back until get at least two matched point
         3.1.1.3 Set the points that we get from the previous step as a datum boundary
      3.1.2. End If loop
   3.2. End If loop
4. End of while loop

The symbol for the GD&T used in the IGES format is shown in Table 7.1.

Algorithm for Test Extraction

1. Open IGES file
2. Create PSection file (store PEntry objects)
3. While Read line from PSection file
   3.1. If there are more than 4 line of 110 PEntery followed by 212 line
      3.1.1. If the last value contains 1H
         3.1.1.1. Read the character after 1H set his value from library as shown in the following table
            3.1.1.1.1. Take it as test name
         3.1.1.2. While read line 212 and if we do not get before this line more than 3 times 110 line

3.1.1.2.1. Extract the value of the test which will be in the last file
3.1.1.3. End of while loop
3.1.2. End of If loop
3.2. End of If loop
4. End of while loop

The general note, which translates the GD&T in IGES file format in the parameter data of the IGES structure, is explained in Figure 7.8.

The output of feature extraction and recognition and GD&T from IGES file format is shown in Figures 7.9 and 7.10, respectively.

Table 7.1 Library of General Note of GD&T Symbol in IGES File Format

IGES Symbol	*Tolerance Type*	*Test Symbol*	*Test Type*
Hu	Tolerance, single element	—	Straightness
Hc		▱	Flatness
He		○	Circularity
Hg		⌭	Cylindricity
Hk		⌒	For line shape
Hd		⌓	For surface shape
Hf	Tolerance guidance for other elements	//	Parallelism
Hb		⊥	Perpendicularity
Ha		∠	Inclination (slope)
Hj	Tolerance of position relative to other elements	⌖	Position
Hr		◎	Concentricity
Hi		⌯	Symmetry

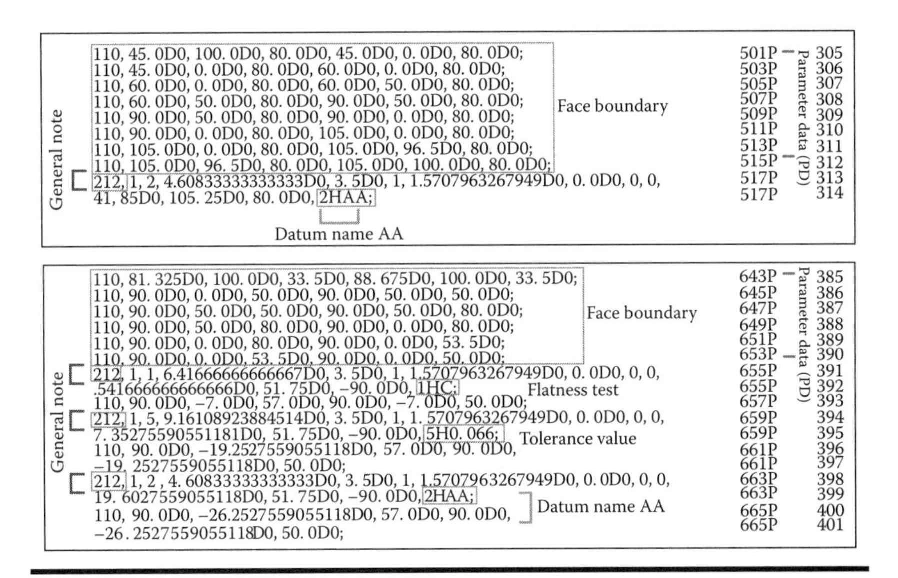

Figure 7.8 Structure of the GD&T in the IGES file format.

Feature id: 1
Type = prismatic (Raw material)
Dimensions:
Length = 80 Width = 100 Height = 120

Feature id: 2
Face1 id: 2 Face2 id: 3
Common edge id: 5
Feature origin: Vertex id: 6 (20, 0, 170.5)
Type: STEP through
Length: 100
Width: 20
Height: 12.5
Volume: 25000

Figure 7.9 Output of feature extraction and recognition from IGES file format.

7.2.5.2 Object-Oriented Programming for Extraction of GD&T from IGES File

The first step toward automatic GD&T extraction is achieved by extracting the GD&T information from the (IGES) CAD file and redefining it as a new object-oriented data structure as demonstrated in Figure 7.11. In this diagram, the IGES class

----

Datum "AA"

Boundary:

"(105, 100, 80)(105, 0, 80)(90, 0, 80)(90, 50, 80)(60, 50, 80)(60, 0, 80)(45, 0, 80)(45, 100, 80)"

Test "Flatness"

"0.066"

"(90, 0, 50)(90, 0, 80)(90, 50, 80)(90, 50, 50)"

Datum "AA"

Figure 7.10 Output of GD&T from IGES file format.

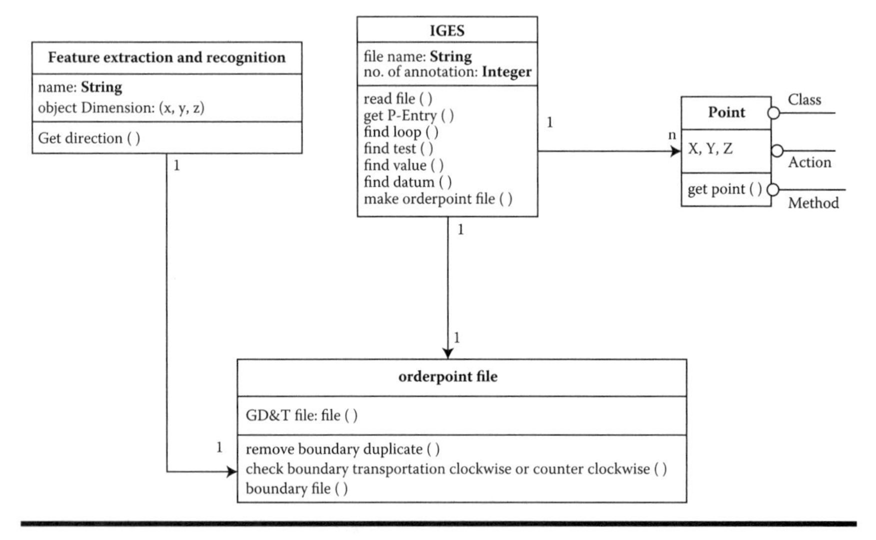

Figure 7.11 OOP class diagram for the extraction of GD&T in IGES file.

is the number of annotation. The IGES class consists of get P-Entry character, get test name, get test boundary, get test type, get test value, get test datum, and make order_point (inspection) file. The point is used for these methods; getting the boundary in the IGES file by getting points (x, y, z). The feature extraction and recognition class consist of these methods to get the direction of the GD&T because it is not supported as the general note in the IGES file. The order_point class contains these methods, removes the duplicate boundary, and checks boundary point's transportation (clockwise or counter clockwise) by reading the boundary file.

7.2.5.3 GD&T Extraction in STEP File Format

The STEP is one of the formats that attaches the different tolerance information on the various features of the part. The STEP generally associates the tolerance entities with SHAPE_ASPECT in order to identify the toleranced feature. The feature of the part in the cases of the solid boundary representation model is mainly represented by ADVANCED_FACE entities in STEP. For example, a through hole in a solid model can be represented by the ADVANCED_FACE entity such as the semicircular surface. The SHAPE_REPRESENTATION, which can be defined as the representation of the SHAPE_ASPECT for the feature, is typically a CONNECTED_FACE_SET and exhibits the same GEOMETRIC_REPRESENTATION_CONTEXT as the solid. The TOPOLOGICAL_REPRESENTATION_ITEMS are collected together by a SHAPE_REPRESENTATION in STEP as shown in Figure 7.12.

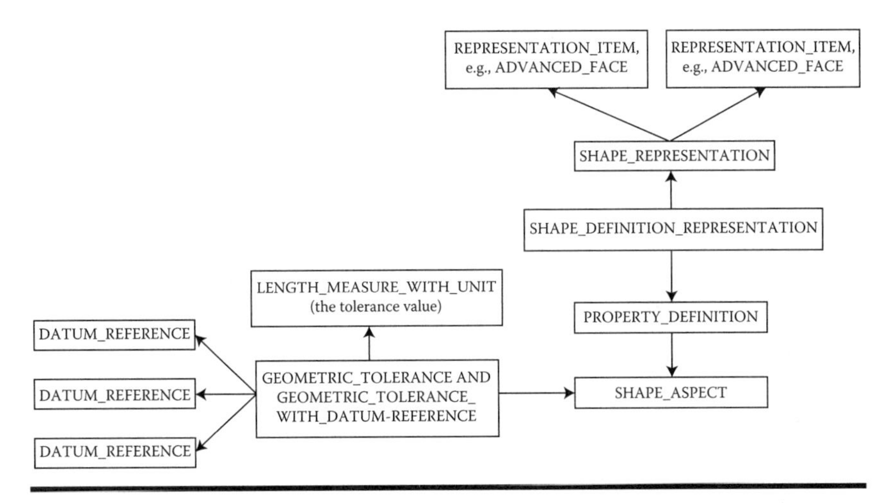

Figure 7.12 Relationship of the tolerance entities and the shape elements.

The application of a COMPOSITE_SHAPE_ASPECT is used rather than SHAPE_ASPECT in the cases where tolerance needs to be applied to more than one feature such as a pattern of holes.

- *Tolerance with datum*: The datum system is used to provide the reference system for describing requirements on the product shape. The need for the datum systems arises when the tolerances are referred to one or more data. The datum systems are defined by datum entities and their corresponding DATUM_FEATURE entities.
- *Datum*: A datum can be defined as a geometric reference such as a point, axis, or plane to which toleranced features can be related. A datum is the origin (or the zero point) from which the location or geometric characteristics of the other features on the part can be established. Since the datum is intended to be the idealized geometry, the unbounded geometric entities are used as the REPRESENTATION_ITEM. For a boundary representation solid model, the geometric entities generally include planes and lines.
- *Datums features*: Datum features are the tangible features on the part such as the face, which provides a reference system for measurements of the actual part. The most important requirement of the datum features is that it must lie on the physical boundary of the shape. Consequentially, DATUM_FEATURE entities are related to topological entities that represent those boundaries in the solid model such as an ADVANCED_FACE. The tolerance entity is considered a subtype of GEOMETRIC_TOLERANCE_WITH_DATUM_REFERENCE where tolerances contain references to data. The block diagram shown in Figure 7.13 explains how the tolerance entity, for example, a POSITIONAL_TOLERANCE using three DATUM_REFERENCE entities can be applied to describe the datum system for the tolerance. The DATUM_REFERENCE points to a datum entity, which, in turn, points to the geometric elements that represent the datum. The DATUM_FEATURE, that is, the feature on the

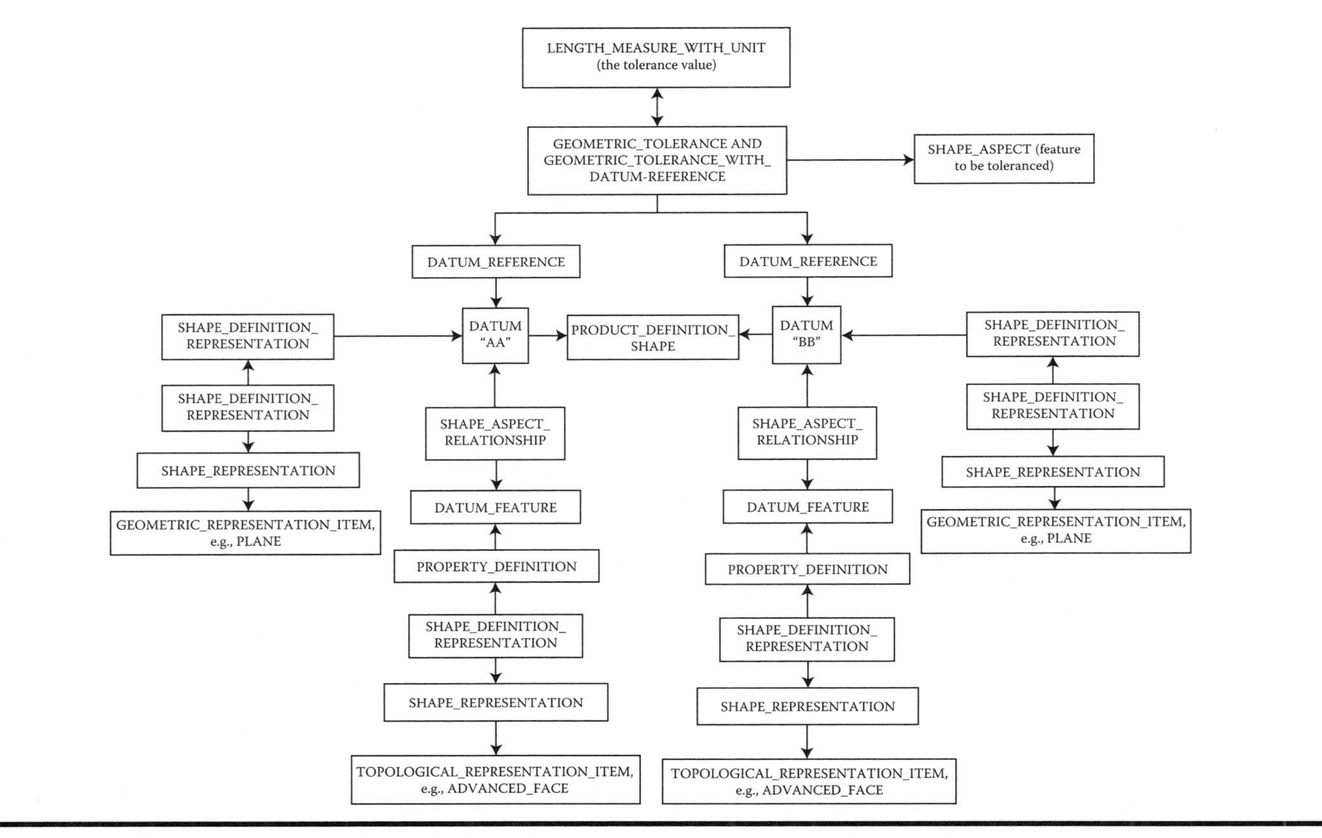

Figure 7.13 Representation of the datums, DATUM_FEATURES, and the datum system.

part corresponding to the datum, references the topological elements of the solid model representing that feature, that is, the ADVANCED_FACE entities.

Algorithm for the Extraction of GD&T in STEP File Format

1. Read STEP file
2. Read lines until lines contain "VALUE_REPRESENTATION_ITEM ('number of annotations')"
3. Find number of annotation N
4. For J = 1 to N
   - 4.1. Give name (J)
     - 4.1.1. Open STEP file
     - 4.1.2. Read lines until line contain "DRAUGHTING_MODEL_ITEM_ASSOCIATION"
     - 4.1.3. Find the name in the line
     - 4.1.4. Print Name
   - 4.2. Read boundary (J)
     - 4.2.1. Open STEP file
     - 4.2.2. Read lines until line contains "GEOMETRIC_ITEM_SPECIFIC_USAGE"
     - 4.2.3. Find the "ADVANCED_FACE"
     - 4.2.4. Use "ADVANCED_FACE" to find face outer
     - 4.2.5. Use "ADVANCED_FACE" to find edge loop
   - 4.3. Read type (J)
     - 4.3.1. Open STEP file
     - 4.3.2. Read lines until line contain "GEOMETRIC_CURVE_SET"
     - 4.3.3. Find type
     - 4.3.4. Print type
5. End

The flow chart describing the algorithm to extract GD&T extraction from STEP file is shown in Figure 7.14.

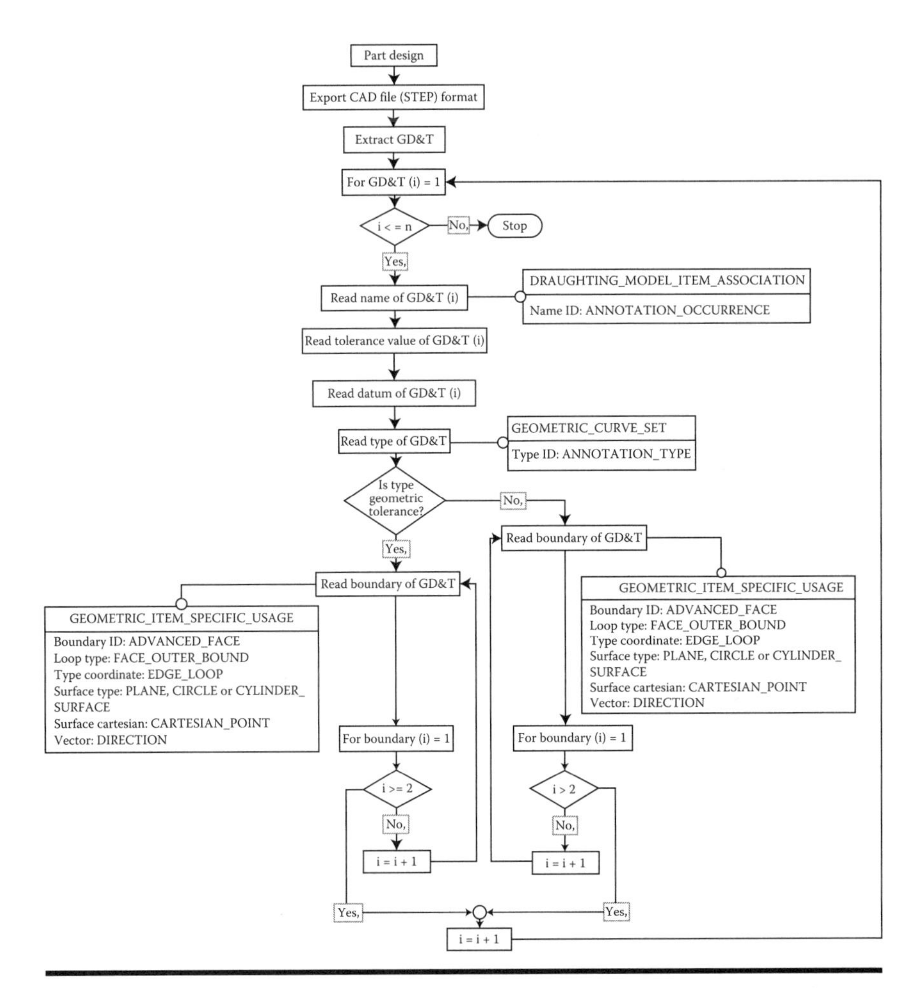

Figure 7.14 Geometrical and dimensional tolerance extraction from STEP file.

7.2.5.4 Object-Oriented Programming for Extraction GD&T from STEP File

The automatic extraction of GD&T information from the (STEP) CAD file and redefining it as a new object-oriented data structure is demonstrated in Figure 7.15.

In this diagram, the STEP class is the number of annotation. The STEP class consists of these methods, get number of annotation, extract GD&T, and make order_point (inspection)

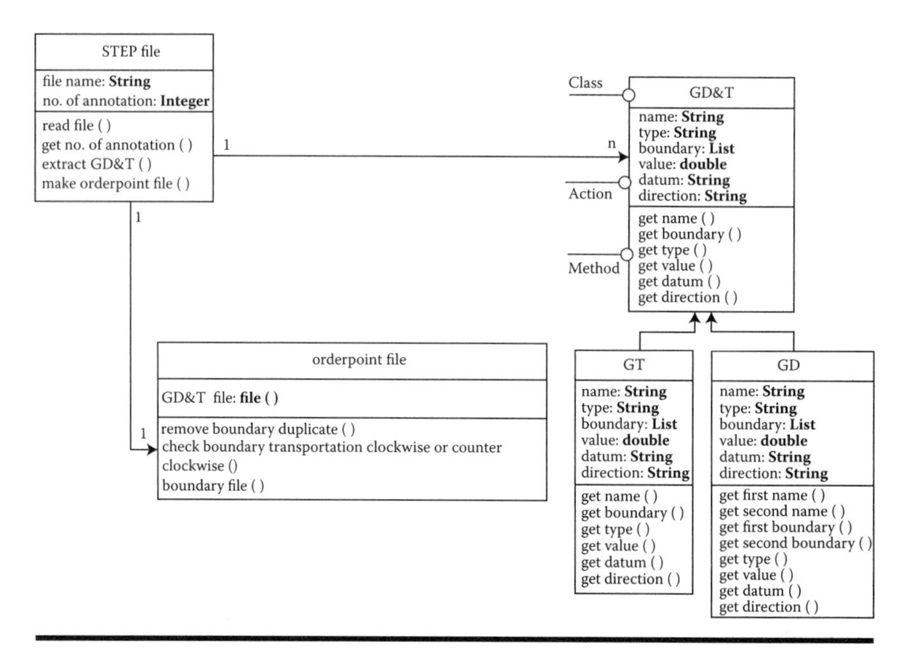

Figure 7.15 OOP class diagram for the extraction GD&T in STEP file.

file. The GD&T class is used for extracting GD&T, which consists of these methods, get test name, get test boundary, get test type, get test value, get test datum, and get test direction. The order_point class contains these methods, removes duplicate boundary, and checks boundary point's transportation (clockwise or counter clockwise) by reading the boundary file.

7.3 Summary

In this chapter, the importance of automatic feature extraction methodology in CAD and CAM integration has been highlighted. Moreover, it includes the following discussions:

- Automatic feature extraction (AFE)
- The different steps of AFE
- Algorithms used for the extraction of machining features
- Intelligent feature recognition methodology

- Feature extraction and recognition using IGES and STEP
- G&DT extraction using IGES and STEP
- The structure of the STEP AP-203
- Hierarchy of form feature

QUESTIONS

1. What is the primary link for the integration of CAD ad CAM systems?
2. Why is CAD and CAM integration important to the manufacturing industry?
3. What is the difference between CAD and CAPP databases?
4. What do you mean by automatic feature extraction (AFE)?
5. Explain the different steps of AFE using a schematic diagram.
6. Discuss the following algorithms used for extracting the machining features from the CAD file:
   a. Syntactic pattern recognition approach
   b. Logic-based approach
   c. Graph-based approach
   d. Section techniques
   e. Volume decomposition and composition approach
   f. Convex-hull algorithm
   g. Cell decomposition
   h. Expert system approach
7. Write down the different steps that can be used for feature extraction and classification.
8. How can the IGES file export GD&T?
9. Explain the different phases of the intelligent feature recognition methodology (IFRM).
10. How are the extracted geometric and topological information from the (IGES/STEP) CAD file defined object-oriented data structures? Explain with a diagram.
11. Briefly explain the structure of the STEP AP-203.
12. Write down the algorithms for single and multiple depression features.

13. Discuss the hierarchy of form features.
14. Write down the algorithm for the extraction of the slot blind feature.
15. Write down the algorithm for datum and test extraction from IGES file.
16. Discuss the output of feature extraction and recognition, GD&T from the IGES file.
17. Describe the OOP class diagram for the extraction of GD&T in the IGES file.
18. What is the difference between datum and datum features?
19. Describe the OOP class diagram for the extraction GD&T in the STEP file.
20. Write down the algorithm for the extraction of GD&T in the STEP file format.

References

1. Y. J. Tseng and S. B. Joshi, Recognizing of interacting rotational and prismatic machining features from 3D mill-turn parts. *International Journal of Production Research*, 1998;36(11):3147–3165.
2. Y. Yildiz, İ. Korkut, and U. Şeker, Development of a feature based CAM system for rotational parts. *G.U. Journal of Science*, 2006;19(1):35–40.
3. S.-C. Liu, M. Gonzalez, and J.-G. Chen, Development of an automatic part feature extraction and classification system taking CAD data as input, *Computers in Industry*, 1996;29:137–150.
4. J. J. Cunningham and J. R. Dixon, Designing with features: The origin of features, *Computers in Engineering Conference*, CIE, San Francisco CA, 1988.
5. M. P. Bhandarkar and R. Nag, STEP-based feature extraction from STEP geometry for Agile Manufacturing. *Computers in Industry*, 2000;41(1):3–24.
6. E. A. Nasr and A. K. Kamrani, *Computer-Based Design and Manufacturing: An Information-Based Approach*, Berlin: Springer, 2007.

7. S. M. Staley, M. R. Henderson, and D. C. Anderson, Using syntactic pattern recognition to extract feature information from a solid geometric database. *Computer in Mechanical Engineering*, 1983;2(2):61–66.
8. S. Joshi and T. C. Chang, Graph-based heuristics for recognition of machined features from a 3D solid model. *Computer Aided Design*, 1988;20(2):58–66.
9. B. Nasser, P. Hamid, and H. R. Leep, A prototype solid modeling based automated process planning system. *Computers in Engineering*, 1996;31(1/2):169–172.
10. O. W. Salomons et al., Review of research in feature based design. *Journal of Manufacturing Systems*, 1993;12(2):113–132.
11. T. C. Woo, Feature extraction by volume decomposition, *Proceedings of the Conference on CAD/CAM Technology in Mechanical Engineering*, Cambridge, MA, 1982.
12. G. F. Luger and W. A. *Stubblefield, Artificial Intelligence and the Design of Expert Systems*, Menlo Park, CA: Benjamin/Cummings, 1989.
13. H. P. Wang and R. A. Wysk, An expert system of machining data selection. *Computer and Industrial Engineering Journal*, 1986;10(2):99–107.
14. International Organization for Standardization, ISO 10303-224. Industrial Automation Systems and Integration-Product Data Representation and Exchange—Application Protocol: Mechanical Product Definition for Process Planning Using Machining Features, 2000.
15. International Organization for Standardization, ISO 10303—Part 11—*Descriptive Methods: The EXPRESS Language Reference Manual*, 1997.
16. S. V. B. Sathi and P. V. M. Rao, 2009, STEP to DMIS: Automated generation of inspection plans from CAD data, *5th Annual IEEE Conference on Automation Science and Engineering*, Bangalore, India, August 22–25.
17. E. A. Nasr and A. K. Kamrani, A new methodology for extracting manufacturing features from CAD system. *International Journal of Computer and Industrial Engineering*, 2006;1(51):389–415.
18. T.-C. Chang and R. A. Wysk, Integrated CAD and CAM through automated process planning. *International Journal of Production Research*, 1984;1(1):158–172.

Chapter 8

Integration System for CAD and Inspection Planning

8.1 Introduction

The geometrical and dimensional inspection of industrial components to ensure conformity to design criteria constitutes an integral part of the manufacturing system. The dimensional inspection planning should be capable of determining plans and information for measuring the dimensions and tolerances of the manufacturing products [1]. In fact, the product quality data, such as geometrical dimensions, form and positional tolerances, are very important for the analysis, evaluation, and decision making. Moreover, an inspection plan defines the measurement and inspection requirements for the manufactured components. However, the inspection process has to be very efficient and provides enough data to determine if the manufactured components satisfy to the design specifications or not. This requires the development of an automated inspection system, which can integrate computer-aided design (CAD), computer-aided inspection planning (CAIP), and coordinate measuring machine (CMM) systems. The design information from the CAD provides input for the generation of CAIP. The inspection procedure developed by CAIP is then transformed into a measuring program for execution on the CMM [2].

The automation of the inspection procedure, which can provide seamless information flow through CAD, CAIP, and CMM system, has been of utmost importance. In fact, the automation of the inspection system comprises the integration of CAD, CAIP, and the CMM systems. The integration between CAD and CAIP assists in the inspection planning task to direct the operation of the inspection system, CMM, for dimensional verification of the industrial components [3]. The goal of the automated inspection planning includes the generation of a program, which can drive the CMM for the measurement of the manufactured part. The objective behind the development of an integrated system is the standardization of the inspection procedure. The integrated system provides the ability to automate the manufacturing activities that currently rely on human interface. This system can contribute several benefits such as reduction of the product lifecycle, improving the product quality, and reducing the manufacturing costs. Meanwhile, the framework of the integrated system (Figure 8.1) is, primarily, composed of three main modules: Automatic features extraction module (AFEM), computer-aided inspection planning module (CAIPM), and coordinate measuring machine module (CMMM).

The integration system can be divided into three major modules: CAD module, CAIP module, and CMM module, which have been described in the subsequent sections in this chapter.

8.2 Development of Computer-Aided Inspection Planning Module

The computer-aided inspection planning (CAIP) module is made up of several components, including set up planning,

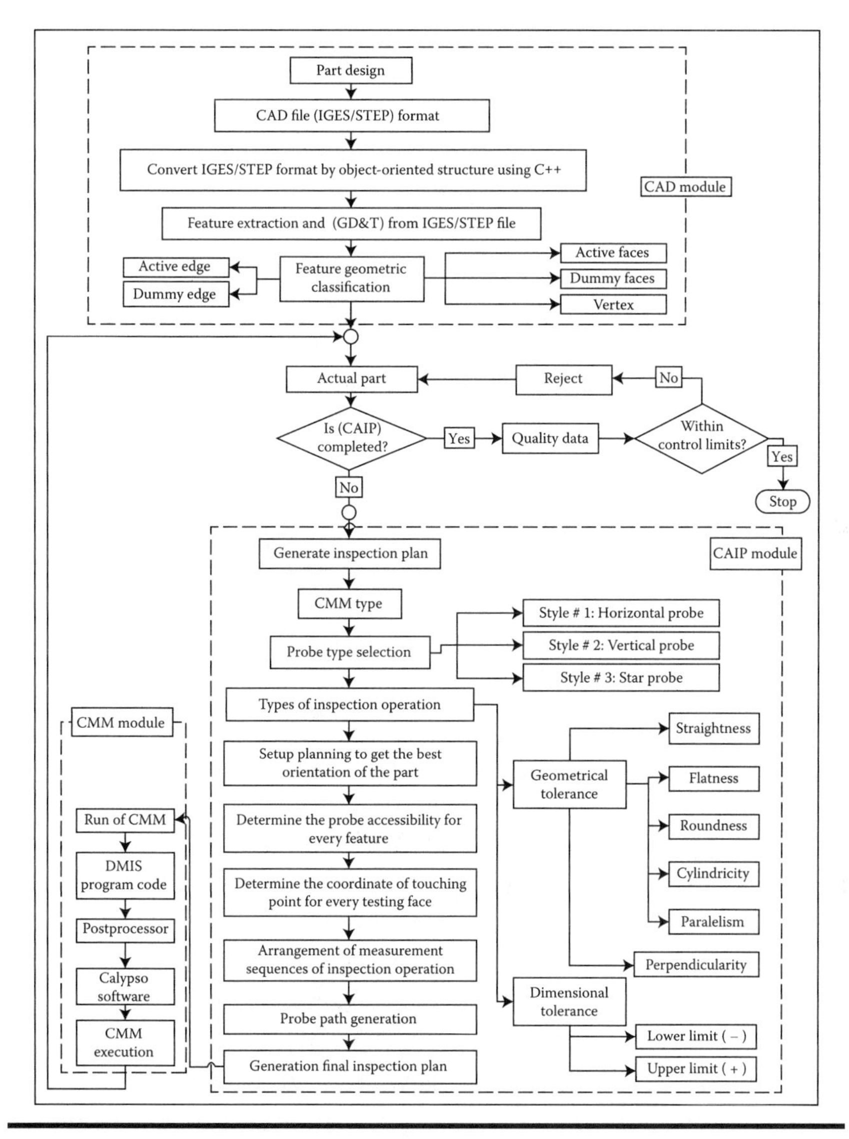

Figure 8.1 Overview of the integrated CAD, CAIP, and CMM systems. (Adapted from A. Kamrani et al., *International Journal of Advanced Manufacturing Technology*, 2015;76:2159–2183.)

probe selection and orientation, accessibility analysis, touch points, etc. to generate an inspection plan [4] as shown in Figure 8.2.

The geometric information extracted from the CAD model using the feature extraction module is used to create the

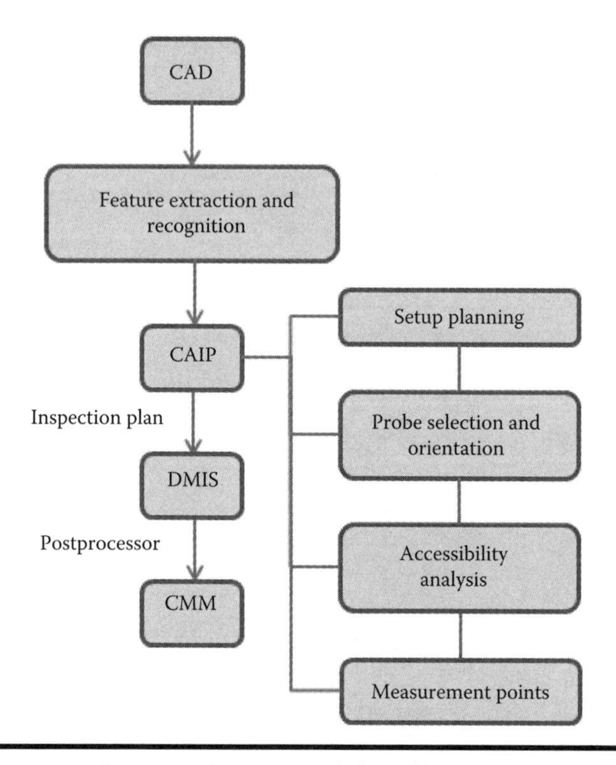

Figure 8.2 Constituents of CAIP module with its input and output.

inspection plan based on the inspection knowledge and rules stored in the system. The inspection planning module generates the inspection attributes, measurement points, part setup plan, and probe approach direction (PAD). The measurement points generated on the CAD model database can be considered as the target point, which the CMM is supposed to touch. Finally, the part setup planning and the PAD are used for selecting the best bottom face of the designed part and directing the probe tip toward the surface of the designed part, respectively. The CAIP module generates the inspection plans and comprehensive instructions for the measurement of the final manufactured component. The integrated system uses a common database that connects and integrates their data. In inspection planning, the approach to measuring designed part attributes that contain both high-level features such as slots, pockets, and holes and low-level entities such as edges and vertices can be realized by taking into consideration the relationships between the extracted features. The input to the CAIP module is obtained from AFEM through the following procedures:

- Using rule, structure, and pointer-based representations of knowledge based on an object-oriented approach
- Developing data models and interfaces to establish the integrating system
- Developing a knowledge-based geometric reasoning approach for automating inspection planning

CAIP module, in fact, develops an intelligent interface between the feature extraction system and inspection planning system. This module contains a database about the information of the designed part, which is required for the generation of the inspection plan.

8.2.1 Module Database

To automate and integrate CAD and CAIP module functions, a structured database is designed as shown in Figure 8.3. The eight elements are generated by the CAD system: object, feature, face, edge, starting vertex, terminating vertex, coordinate, and inspection specifications [5]. The object properties start from a designed part that is linked to its extracted feature group. Each feature is linked to its face group, which is linked to its group of edges in the edge group, and which, in turn, are linked to its group of vertices in the starting and terminating vertex group. At the end, each feature is linked to its inspection specification group. The database can be extended to include three groups: criteria group, setting group, and dummy active group, which are used as the main data to generate the inspection planning instructions.

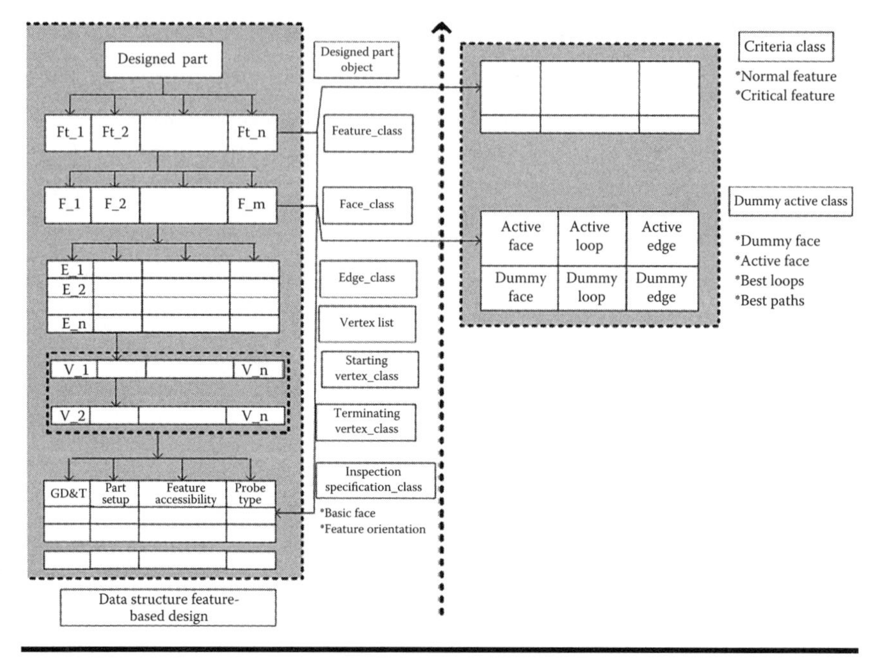

Figure 8.3 The data structure of the CAIP module.

8.2.2 Developing the Integration between CAD and CAI

The CAD system database is made up of low-level entities such as points and lines of a designed part. This geometric information is not suitable for downstream applications such as process planning and inspection planning. Therefore, the feature extraction approach is utilized to recognize the design of part geometry, tool geometry, tool motion path, probe motion path, and features that can provide all the necessary information for process planning and inspection planning. Hence, a computer-integrated manufacturing environment can be achieved by creating a good database of the designed part.

The manufacturing features can be categorized into simple features and complex features. The complex features are created by the interaction of two or more simple features. The basic element of any feature is the face on which the feature is created. This face can be identified as the basic face (BF). Once the detailed database of the extracted features from CAD system is established, the inspection plan and their sequences can be identified. The feature extraction module plays an important link in the integration of CAD, CAIP, and CMM systems.

8.2.3 Generation of the Inspection Plan for the Manufactured Components

The inspection activities and their sequences can be determined from the feature extraction module using CAD information. The main problem in the integration of CAD and CAIP modules is the automatic and direct design data interpretation without the need of the extraction of technological and manufacturing information. However, the feature extraction methodology introduced for this work eliminates the need of an interface between the two activities. It also provides an efficient processing of data knowledge of the design stage in order to support the downstream activities.

The inspection planning module is a knowledge-based system, as shown in Figure 8.4. The preparation of inspection knowledge entails the following items:

1. Working faces (BF) on which features are created
2. High-level feature types created on a certain (BF)
3. Feature orientation and probe location (S)
4. Setting for each feature, which is determined by feature type and orientation
5. Inspection parameters and measuring edges of a feature for each setting
6. Edge limits and edge values
7. Probing parameters (i.e., probe approach direction and probe approach sequence)

A major component of the inspection planning module is the inspection database, which includes declarative database and procedural database. The declarative database consists of

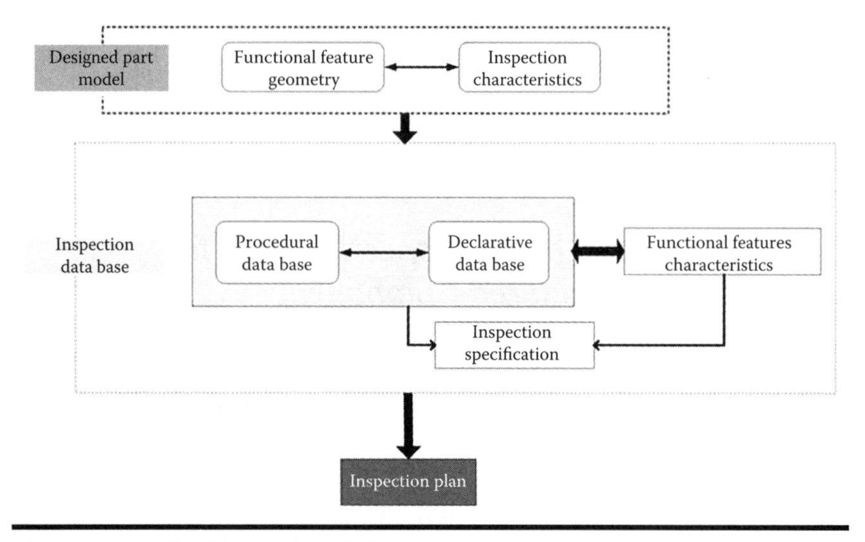

Figure 8.4 The hierarchy of the inspection module.

information about the part and its features, inspection characteristics, specifications, and manufacturing processes. In fact, both the declarative and procedural databases can be considered as a system's problem-solving knowledge [5].

8.2.3.1 Feature Classification in the Inspection Plan Generation

The geometric features have to be classified in the CAIP module in order to prepare for the different measurement steps. Although the measuring component is defined by the single feature of the CAD/CAM, still, for the detailed analysis, the measuring featuring has to be segregated into surface information. The measuring geometries can be categorized into free-form geometries and analytic geometries. The analytic geometries can further be subdivided into single and compound geometries. In a single geometry, the object can be divided into three entities: (a) point, (b) line, which can be classified as a straight line and curved line, and (c) surface, which can be classified into the plane, cylinder, cone, and sphere. Moreover, the detailed inspection of any given surface, at least, requires the determination of six geometric tolerances, including position, straightness, flatness roundness, conicity, and cylindricity. The inspection of the integrated geometries (two or more single geometries) requires the analysis of the four geometric tolerances including parallelism, squareness, angularity, and concentricity [6]. The inspection plan for CMM can broadly be divided into high-level planning and low-level planning [7]. For example, the tasks such as workpiece setup, probe selection, and orientation constitutes a high-level planning while the selection of measuring points, probe path generation, and probe execution are defined as the low-level planning. Similarly, Lee et al. [8] developed a CAIP system made up of two stages, including global inspection planning and local inspection planning for the generation of the inspection plan [8]. The global inspection plan consisted of activities such as the sequence of setups and form features for the part while determination and sequencing of measuring points on the surface features constituted local inspection planning.

8.2.3.2 Accessibility Analysis

The accessibility analysis of the measuring component is very crucial in the generation of the inspection plan table [9]. It significantly reduces the number of unnecessary changes in probe orientation and maximizes the number of features inspected using the same probe orientation [4].

8.2.3.2.1 Probe Accessibility Direction

The probe accessibility direction (PAD) represents the accessibility direction of the probe as it makes contact with each feature and the direction of the individual or the group features. In fact, it can be defined as the direction of approach of the probe toward the inspecting feature. The PADs for any feature or the component can be determined by applying the clustering algorithm. The implementation of the clustering algorithm requires the grouping of all the features having same PAD and

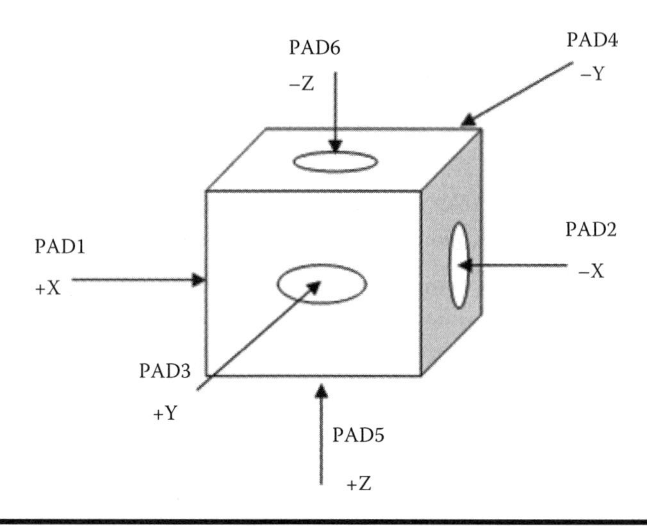

Figure 8.5 Probe accessibility direction (PAD).

which can be inspected in one operation. The PADs for the different features are shown in Figure 8.5.

After analyzing the feasibility of PAD for each feature, the features of a PAD matrix can be constructed. Being a part with m features that can be inspected using n PADs based on the inspection probe, PAD matrix is a m × n matrix as shown in Figure 8.6.

Where

m is the number of rows in the PAD matrix, that is, number of features

n is the number of columns in the PAD matrix, that is, number of probe orientations

If feature f_j has PAD_i, then $r_{ij} = 1$, otherwise $r_{ij} = 0$

	PAD1	PAD2	PAD3	PAD4	PAD5	....	PADn
f1	r11	r12	r13	r14	r15	....	r1n
f2	r21	r22	r23	r24	r25	....	r2n
f3	r31	r32	r33	r34	r35	....	r3n
f4	r41	r42	r43	r44	r45	....	r4n
⋮	⋮	⋮	⋮	⋮	⋮	....	⋮
fm	rm1	rm2	rm3	rm4	rm5	....	rmn

Figure 8.6 PAD (m × n) matrix.

8.2.3.2.2 Approach Direction Depth

Approach direction depth (ADD) of a feature can be defined as the depth accessibility of the probe during the inspection [4]. The ADD is measured from the highest point on a part to the lowest point in the feature with the part orientated in PAD. A slot through feature, as shown in Figure 8.7, can be measured using PAD1, PAD2, and PAD3 with ADD of ADD1, ADD2, and ADD3, respectively.

After calculating the feasible ADDs for each feature, a feature ADD matrix can be constructed. For a part with m features that can be inspected using n PADs, an ADD matrix is an m × n matrix as shown in Figure 8.8, where

m is the number of rows in the matrix; that is, number of features in the part

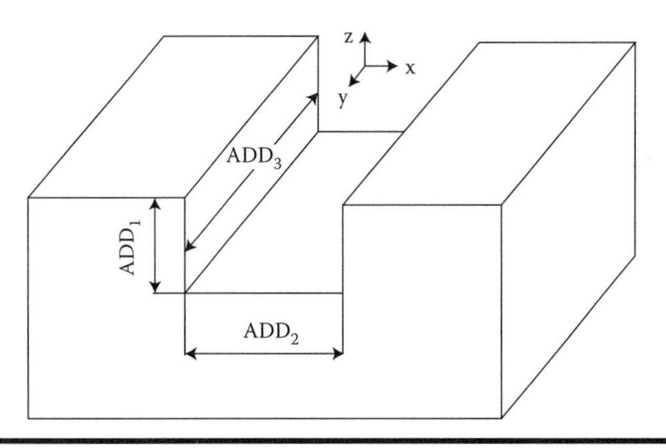

Figure 8.7 Slot through feature.

	PAD1	PAD2	PAD3	PAD4	PAD5	....	PADn
f1	g11	g12	g13	g14	g15	....	g1n
f2	g21	g22	g23	g24	g25	....	g2n
f3	g31	g32	g33	g34	g35	....	g3n
f4	g41	g42	g43	g44	g45	....	g4n
⋮	⋮	⋮	⋮	⋮	⋮	....	⋮
fm	gm1	gm2	gm3	gm4	gm5	....	gmn

Figure 8.8 ADD matrix.

n is the number of columns in the matrix; that is, number of probe orientations

g_{ij} is the ADD of feature f_j

From the PAD matrix, a probe type will be selected depending on the features covered at a given setup. ADD will give some information required in the probe selection, such as the length of the probe and the diameter of the probe sphere. The algorithm for PAD analysis can be described as follows:

- Open setup file
- While file is not the end, then read line by line for each feature
- Read all faces in the feature and normal vector
- Create matrix contains six column (+x, −x, +y, −y, +z, −z)
- Store the first element from normal vector in columns (+x, −x) based on the sign of the element
- Store the second element from normal vector in columns (+y, −y) based on the sign of the element
- Store the third element from normal vector in columns (+z, −z) based on the sign of the element
- Compute the summation of every column in the matrix
- If the summation is greater than zero
  Set it = 0
  Else
  Set it = 1
- End if loop

For example, a blind slot with dimensions of 20, 60, and 20 units length, width, and height, respectively, shown in Figure 8.9 has only two probe directions.

The normal vector for each face of the feature has been used to find the accessibility of the probe in the slot blind feature as shown in Table 8.1.

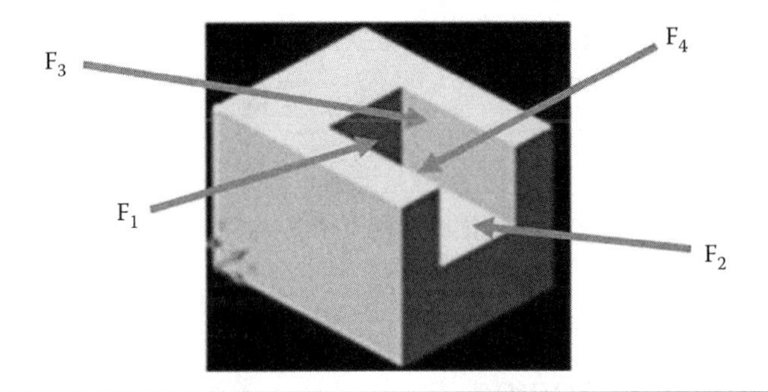

Figure 8.9 Slot bling having only two probe directions.

The PAD matrix of the slot through is as follows:

$$\text{Slot Through}\begin{bmatrix} +x(PAD) & -x(PAD) & +y(PAD) & -y(PAD) & +z(PAD) & -z(PAD) \\ 0 & 0 & 1 & 0 & 0 & 1 \end{bmatrix}$$

The ADD matrix of the slot through is as follows:

$$\text{Slot Through}\begin{bmatrix} +x(PAD) & -x(PAD) & +y(PAD) & -y(PAD) & +z(PAD) & -z(PAD) \\ 0 & 0 & 60 & 0 & 0 & 20 \end{bmatrix}$$

Table 8.1 Normal Vector for Every Face of the Feature

	Normal Vector	*(+X)*	*(−X)*	*(+Y)*	*(−Y)*	*(+Z)*	*(−Z)*
Face Id# 1	(−1,0,0)	0	1	0	0	0	0
Face Id# 2	(1,0,0)	1	0	0	0	0	0
Face Id# 3	(0,−1,0)	0	0	0	1	0	0
Face Id# 4	(0,0,1)	0	0	0	0	1	0
Sum		1	1	0	1	1	0
Accessibility		0	0	1	0	0	1

The PADs for the slot blind feature are +y and −z directions within ADDs 60 and 20, respectively, as shown in the matrices. This information from the PAD and ADD matrices has to be used to select the best probe (considering the length of the probe and diameter of the probe sphere) for a given feature.

8.2.3.3 Setup Planning

The probe accessibility direction (PAD) is an important parameter for the determination of all setups required to measure all the existing part features. Meanwhile, the PAD of a given feature can be defined as an unobstructed path that a probe can take while accessing that feature [10]. The features with the same PAD can be grouped into one setup. Moreover, a feature can have more than one PAD; therefore, it can be grouped into various setups. In the setup planning, the grouping of the features depends on PAD and ADD in order to

- Determine the preferential base for placing the part which allows the probe to access the majority of the features
- Cluster of features based on the common preferential base
- Filter multiple occurrences of the features in different clusters [11]

The setup planning can be defined as a decision-making process, which determines how a given part should be oriented on CMM machine table [12]. A proper setup planning results in the measurement of a maximum number of features with minimum setup. The setup planning becomes important, especially, when the time required to change part setup is quite high with respect to the overall inspection time of the part. The primary objective of the setup planning is to determine the part face that consists of the minimum number of inspecting features. This face determines the base face for part orientation on the machine table. In fact, the base face is called as preferential base or primary locating face in setup planning. Any part face that would result in inspection of maximum possible features should define part orientation for the best setup plan.

8.2.3.3.1 First Rule (Numerical Method)

This method makes use of artificial neural network (ANN) to predict the best setup. The geometric extracting entities and features with same PAD are used as an input to the ANN. The inputs to the ANN can be identified as follows:

- Geometric extracting entities: It comprises geometric entities such as the number of vertices, line edges, circular edges, internal loop, external loop, concave faces, and convex faces. All these entities are obtained from AFEM.
- Features having same PAD: The features with same PAD can be determined as follows:
  - Six PADs (+x, −x, +y, −y, +z, and −z) and six setups (S (right), S (left), S (front), S (rear), S (top), and S (bottom)) are identified in the given rectangular block.
  - For the given setup, the number of features that can be accessed for each PAD are determined and expressed in the form of (m x n) matrix as shown in Figure 8.10. In (m x n) matrix, f11 represents the number of features that can be inspected with +x probe direction

	+x(PAD)	−x(PAD)	+y(PAD)	−y(PAD)	+z(PAD)	−z(PAD)
S(Right)	f11	f12	f13	f14	f15	f16
S(Left)	f21	f22	f23	f24	f25	f26
S(Front)	f31	f32	f33	f34	f35	f36
S(Rear)	f41	f42	f43	f44	f45	f46
S(Top)	f51	f52	f53	f54	f55	f56
S(Bottom)	f61	f62	f63	f64	f65	f66

Figure 8.10 Number of features for different PADs and the part faces.

when the right face of the part is used as the primary locating face. Similarly, f54 represents the number of features that can be inspected with –y direction when the top face acts as the primary locating face.

- For each PAD, the total number of features that can be accessed is calculated, for example, the total number of features for +x PAD is equal to (f11 + f21 + f31 + f41 + f51 + f61). Similarly, the total number of features for +z PAD is equal to (f15 + f25 + f35 + f45 + f55 + f65).
- Finally, the six inputs in form of summation of PADj in each column are obtained.

- Training and testing of experiments: The training experiments determine the number of hidden layers plus optimize the network structure as shown in Figure 8.11. A large number of training experiments with different numbers of hidden neurons, learning rate, and the momentum values (0.80) are checked to obtain the best training parameters and minimum error.
- Result of testing: After several training experiments, the network is successfully trained with some average validating error as shown in Figure 8.12.

  Once the training is successfully finished, the network is validated for a given set of examples and the validating percentage is obtained.

8.2.3.3.2 Second Rule (Graphical Method)

In this method, the best setup is determined using a number of interacting features as shown in Figure 8.13. The face with the minimum number of interactions is selected as the primary locating face.

The different steps that can be used to determine the best setup for any prismatic part can be described as follows:

- Divide all the faces of a prismatic part as primary faces and secondary faces.

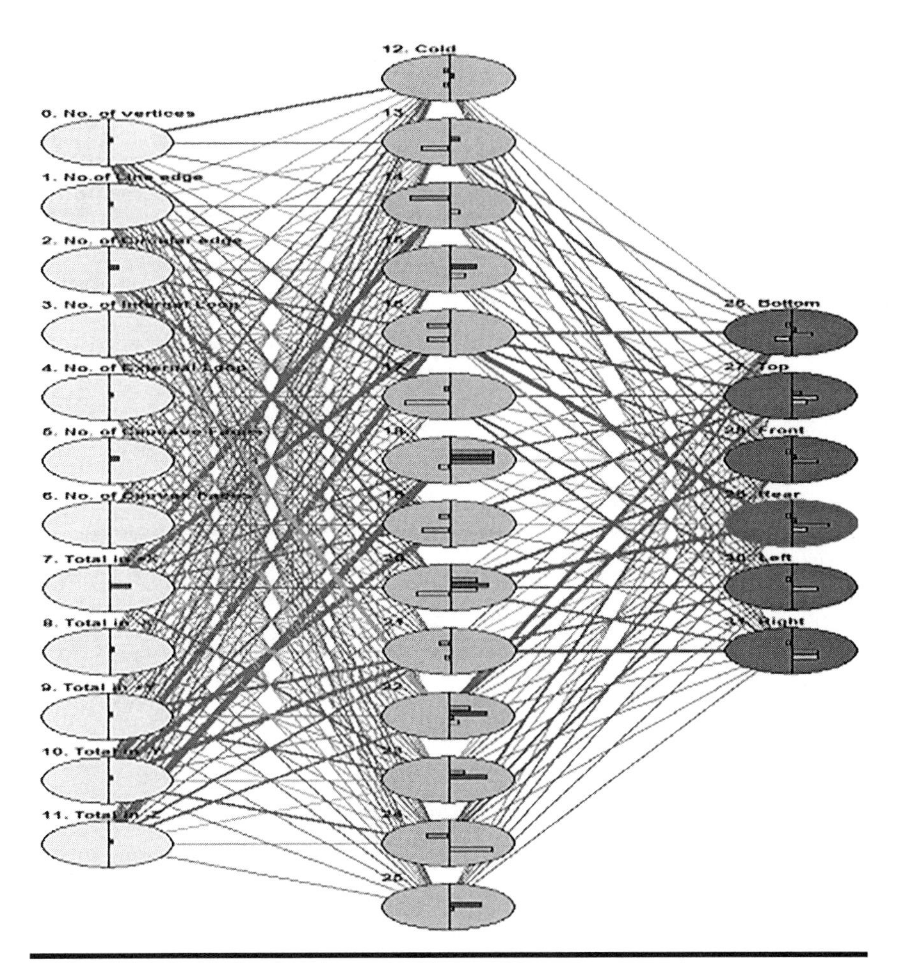

Figure 8.11 Topological structure of back propagation neural network.

The primary faces of the rectangular block are the faces that determine the basic shape of the part. For example, top, bottom, front, rear, right, and left faces fall into this category.

The secondary faces are the faces that belong to the different features on the part such as slot, rib, boss, and pocket.

- Determine interaction between primary faces and edges of secondary faces. For example, left face f24 (as shown in Figure 8.14a) has an interaction with only one edge, top faces f27 and f17 (as shown in Figure 8.14b) have

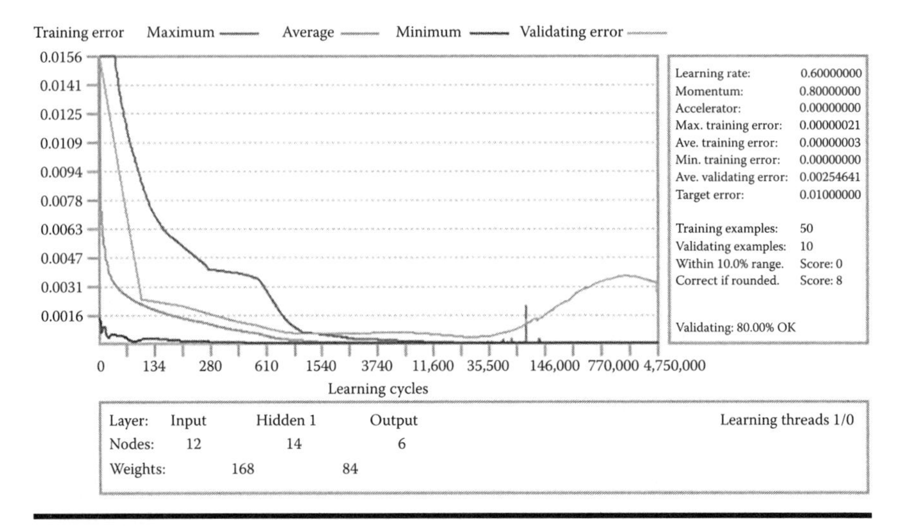

Figure 8.12 Learning cycle—training with 1 hidden layer and 14 nodes.

interactions with eight edges, right face f28 (as shown in Figure 8.14c) has an interaction with only one edge, bottom faces f12 and f21 (as shown in Figure 8.14d) have interactions with six edges, front face f7 (as shown in Figure 8.14e) have interactions with 16 edges and rear face f16 (as shown in Figure 8.14f) have interactions with 10 edges.

The primary faces should be arranged in the ascending order of the number of interactions in order to find the best setup.

- Select the primary face with the minimum number of interactions as the primary locating face. For example, left face f24 and right face f28 have the least interactions (one interaction each). Therefore, either the left face or the right face can be selected for the primary locating face as shown in Figure 8.15. The primary locating face (left face f24 or right face f28) will allow the probe to inspect the maximum number of features in one setup. With this orientation, the probe can inspect features with feature IDs (2, 4, 5, 6, 7, and 8).

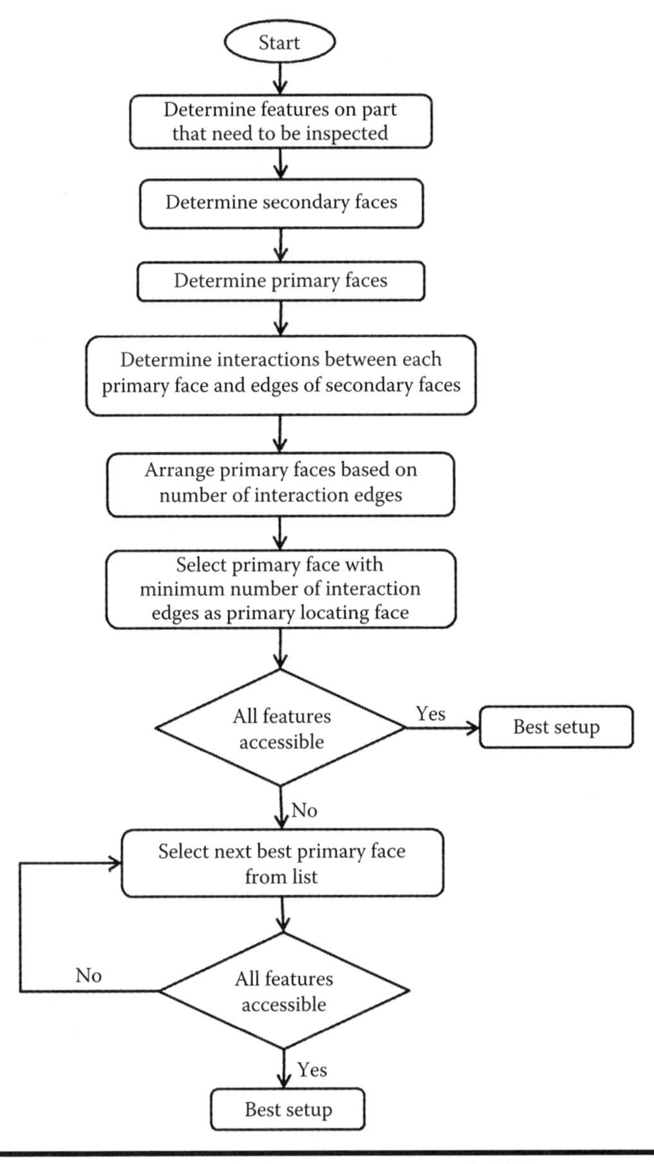

Figure 8.13 Graphical method—setup planning algorithm.

8.2.3.4 Touch Point Generation

For the inspection of any part using CMM, the numbers of measurement points and their coordinates of the given surface have to be determined. When the large number of points is selected, the accuracy of the measurement results

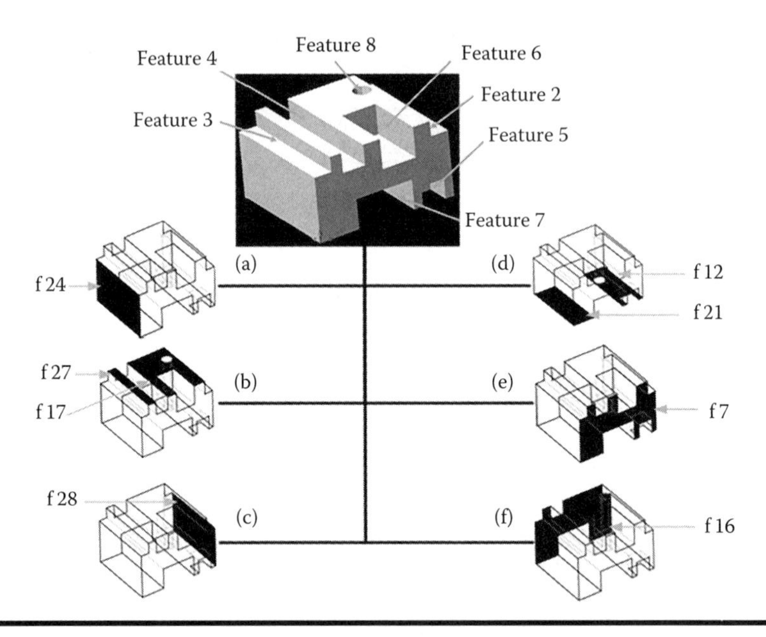

Figure 8.14 Prismatic part and interactions edges. Primary face—(a) Left face (f24); (b) top face (f17 and f27); (c) right face (f28); (d) bottom face (f12 and f22); (e) front face (f7); and (f) rear face (f16).

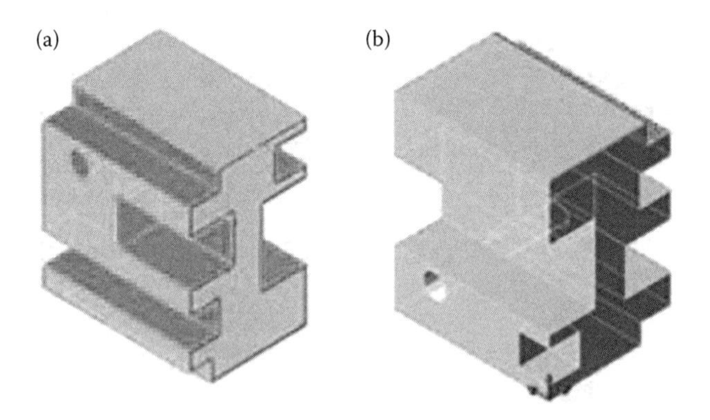

Figure 8.15 The best setup using graphical method. (a) Left as primary locating face and (b) right face as primary locating face.

increases, but the measurement time also increases. In real industrial applications, the determination of an adequate number of points for a given feature is very crucial [13]. The different operators can have different choices based on their experiences. The determination of an optimum number of measuring points for any given feature is a critical task. The determination of the number of points for any surface depends on the type of face geometry and the type of geometrical tolerance on the face [14]. There are different algorithms that can be used to determine the number of probing points. There are basically three types of surfaces, namely, rectangular, cylindrical, and conical face.

The measuring points should be a good representative of the entire surface, the coordinates (or the location) of the points should be such that a maximum amount of information can be obtained [15]. These coordinates can be determined using statistical procedures or from the knowledge of manufacturing processes or a combination of both. There are a number of methods, including the uniform sampling, random sampling, stratified sampling, Hammersley sequence sampling, and Harlton Zaremba sampling [13] to determine the location of the measuring points.

8.2.3.5 Probe Path Generation

The probe movement between the measuring point at the given feature and the subsequent selected features can be defined as the probe path. The probe path inside any feature depends on the sequence of the coordinate touch point. There are some basic principles for the path planning [13]:

- To approach the measuring point from the normal direction of the surface where the point exists
- Maintain a clearance distance before probing
- Avoid collision between the probe and the measuring object

The probe collision, as shown in Figure 8.16, occurs when the probe moves between the different features for the purpose of measurement. The collisions can be divided into two categories: the probe collision and the probe holder collision [6]. The probe collision occurs when the measurement of the feature is finished and the probe moves to the next feature. The probe's

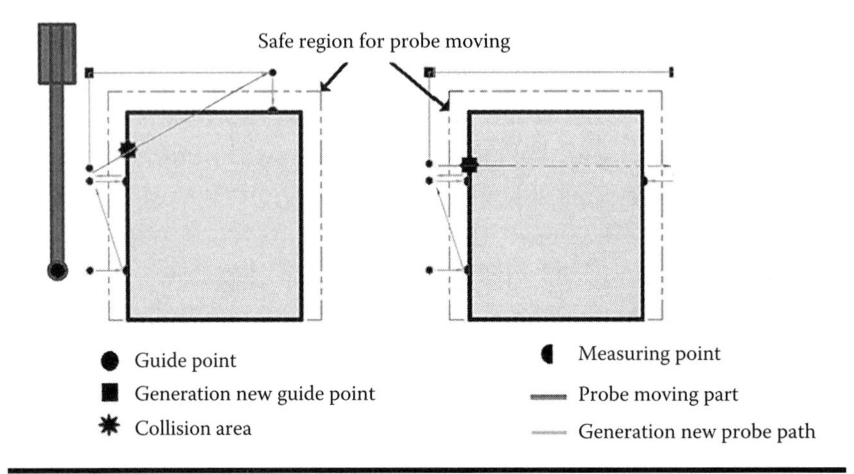

Figure 8.16 Probe collision avoidance.

holder collisions occur when the probe's holder makes contact with the surface of the part as shown in Figure 8.17. It is primarily affected by the feature accessibilities, dimensions, and the probing point's coordinates inside the feature. It can be avoided by managing the probe dimensions and orientations.

The procedures of the probe avoidance collision can be discussed as follows:

- A last touching point of the previous feature is selected and denoted as last point (Xo, Yo, Zo).

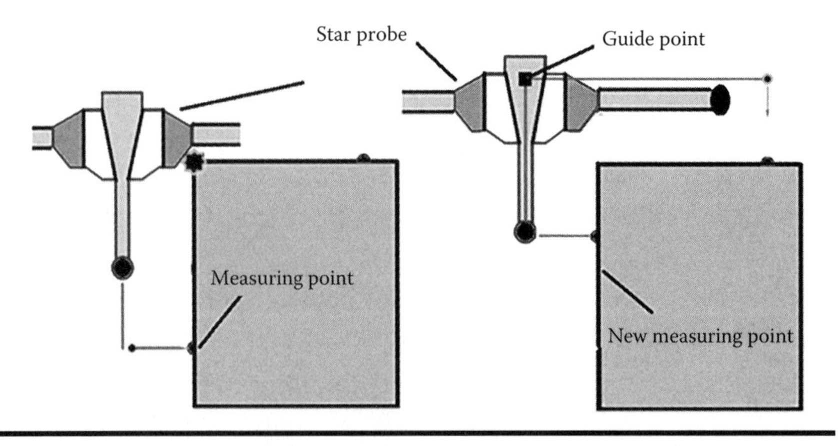

Figure 8.17 Probe collision holder avoidance. (Adapted from M.-W. Cho et al., *KSME International Journal*, 2004;18(8):1349–1357.)

- The first touching point of the next feature is selected and denoted as target point (Xn, Yn, Zn).
- A clearance distance value is entered for moving the probe in the free space of the part (CL and CLz), and this value is small in X- and Y-direction (CL) and large in Z-direction (CLz).
- Check the direction of the previous feature for generating the first guide point of the probe in an opposite direction of its plan.
- Generate the second guide point of the probe by replacing the z cell value of the first guide point with the CLz value.
- Check the direction of the next feature for generating the third guide point of the probe. If the next feature direction is in X- or Y-direction, the CL value is subtracted by the Xn or Yn cell target value, respectively, and replaced from the X or Y cell value of the second guide point. Next, the fourth guide point of the probe is generated by replacing the other X or Y cell of the third guide point with the other Xn or Yn, respectively, of the target point value. Finally, the fifth guide point of the probe is generated by replacing the Z cell value of the fourth point with Zn cell value of the target point.
- Check the direction of the next feature for generating the third guide point of the probe. If the next feature direction is in Z-direction, the X or Y cell of the second point is replaced with Xn or Yn, respectively, of the target point value. Next, the fourth guide point of the probe is generated by replacing the other X or Y cell of the third point with the other Xn or Yn, respectively, of the target point value. Finally, the fifth guide point of the probe is generated by subtracting by the CLz value with Zn cell target value and replaced the result from the z cell value of the fourth guide point.

8.2.4 Inspection Planning Table

The inspection plan table should contain all the information obtained through feature extraction and recognition, accessibility analysis, setup planning, and determination of measuring points as shown in Table 8.2 [12]. This table can be summarized as follows:

1. Face ID is the ID number of the face in the features extraction and recognition output.
2. ID of the inspection operation is the type of the GD&T like flatness and perpendicularity.
3. Tolerance value allows designers to set tolerance limits for all of the various critical characteristics of the part by identifying its function and its relationship to mating parts.
4. Tool used for inspection such as horizontal probe, vertical probe, and star probe depending on the location, depth, dimensions, and orientation of the features and radius, and length of the probe to avoid the collision between the part and the probe.
5. Datum faces are the reference points, lines, planes, and axis, which are assumed to be exact.
6. Orientation of the part is the best orientation of the prismatic part and the number of setups.
7. Number of touch points.
8. Touch points are the coordinates of the probing point.
9. Geometric inspection boundary represents the boundary of the testing face.

8.3 Coordinate Measuring Machine Module

Coordinate measuring machine (CMM) can be defined as the precision measurement equipment connected to a computer with the measurement system software. These measurement

Table 8.2 Inspection Plan Table

Face ID	*ID of Inspection Operation*	*Tolerance Value*	*Tool Used*	*Datum Faces ID*	*Orientation of Part*	*No. of Touch Point*	*Coordinates of Touch Point*	*Geometric Inspection Boundary*
1	G&DT	Tolerance value	Probe Type	Face ID	Setup No.	n	(Ui, Vi, Wi) – (Ui + 1,Vi + 1,Wi + 1) (Ui + 2,Vi + 2,Vi + 2) . . (Un, Vn, Vn)	(X1,Y1,Z1)- (X2,Y2,Z2)- (X3,Y3,Z3)- (X4,Y4,Z4)-

systems can provide very simple inspection as well as a highly complex programming environment. A successful inspection planning methodology is needed even for the measurement of simple manufacturing features using CMM. The inspection planning methodology is based on the critical inspection specifications rather than checking all dimensions of the designed part. Moreover, the module has the ability to select the optimal measuring points and generate a probe path for the CMM. Therefore, by optimal selection of the probing sequence, the probe movement, and inspection time can be reduced. The methodology for the CMMM module can be described in Figure 8.18.

8.3.1 Machine Settings

The machine settings allow the users to set the machine parameters prior to the actual measurement. It usually includes functions for speed, acceleration, etc. The dimensional measuring interface standard (DMIS) provides statements within the command syntax for machine setup within the inspection program. This provides portability of the program, which means that if the program is copied to another machine, it will set the machine up using the program statements, rather than depending on the user of the second machine.

8.3.2 Probe Calibration

The probe calibration is referred to as sensor calibration in DMIS and is required for most machines prior to the use of the probe for measurement.

8.3.3 Datum Alignment

The datum alignment or part setup is needed in order to set the position and orientation of the coordinate system. The part

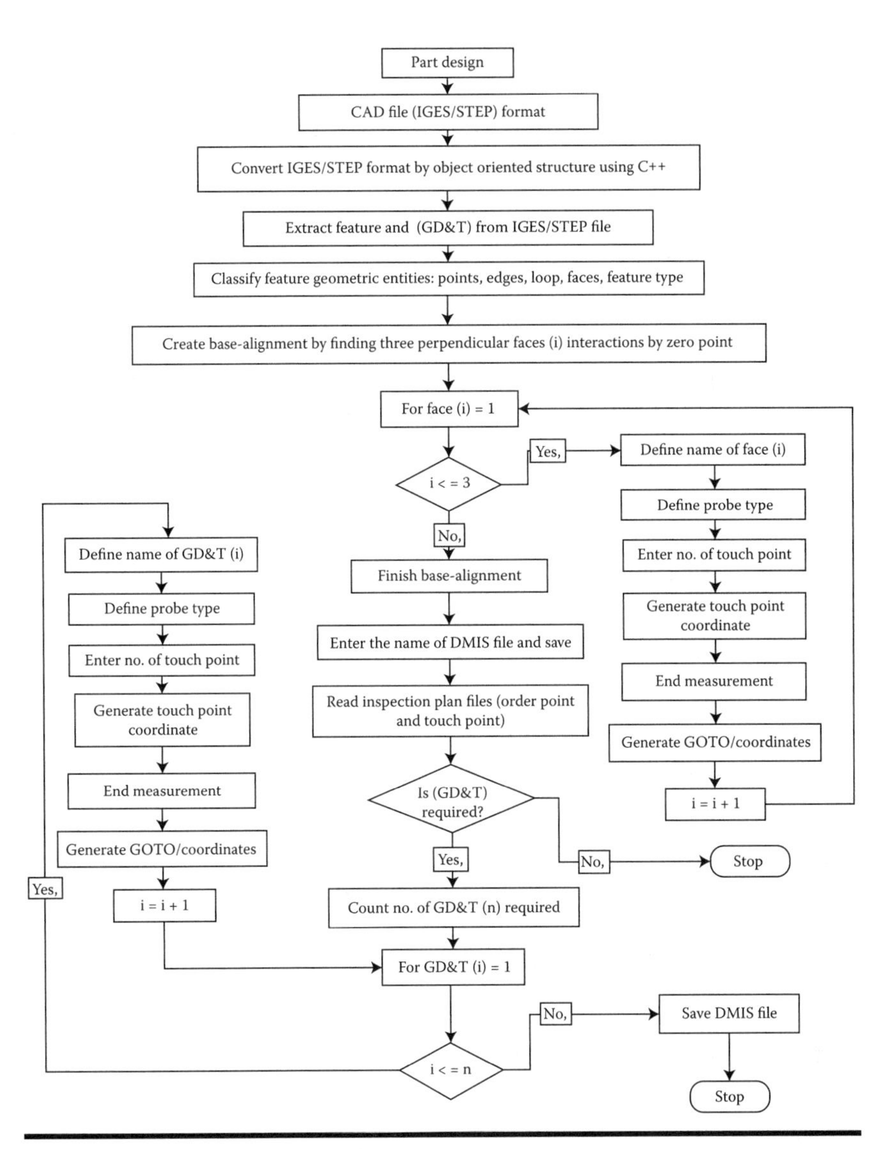

Figure 8.18 CMMM procedure.

setup is usually defined by the geometric tolerance definitions contained in the component drawings.

8.3.4 Measurements

The measurement usually consists of touching a point on the feature of the component being measured.

8.3.5 Output

The output of the results is the reason to all the trouble of setting up a common and measuring it. Some CMM machines output the actual results at the time of measurement.

8.3.6 DMIS File Generation

The DMIS is a standard format of a high-level programming language and used for bidirectional transfer of inspection data between the CAD systems and the CMM. The automation is achieved by creating the tolerance geometric model in a CAD system and automatically generating the DMIS program required for the execution of the inspection plan. The generated program of measuring features is executed and compared to the GD&T like position, orientation, size, and flatness, depending on the allowable tolerances associated with the feature. The automatic generation of the table of the inspection plan is carried out based on the following steps:

- Read the extraction file output and use it as an input to generate the DMIS file automatically.
- Generate the base alignment by finding the zero point (base point of the part), which is obtained by the intersection of the three face.
- After the completion of base-alignment faces, the name of DMIS file is entered and the base-alignment file is closed.
- Check if there are GD&T required or stop.
- If there are GD&T needed, the total will be counted in the For loop in sum (N).
- Enter the name DMIS file and the name of the DMIS base-alignment file.
- Enter the number of touch points needed for the GD&T (i).
- Read GD&T (i) face ID.
- Read the probe type ID depends on the probe accessibility analysis.

- Read the probe path generation between selected features.
- Read GD&T (i) type.
- Read GD&T (i) tolerance value.
- Read datum ID of the GD&T (i) and store it in the fifth column. If there are no datums keep it empty.
- Make the output of the test depending on the reference datum.
- Finish the data required for the GD&T (i).
- Check if there are more GD&T required or stop and close the DMIS file generation.

8.3.7 OOP Class Diagram of DMIS Generation File

The automatic DMIS generation can be achieved by reading the feature extraction and recognition file and redefining it as a new object-oriented data structure as demonstrated in Figure 8.19.

In Figure 8.19, the base-alignment class consists of actions such as importing of feature extraction and recognition file, number of touch points, measure x-distance, y-distance, and z-distance between the home position of the CMM and the actual part. These actions will get the zero point of the CAD model, generate touch point, generate DMIS base-alignment

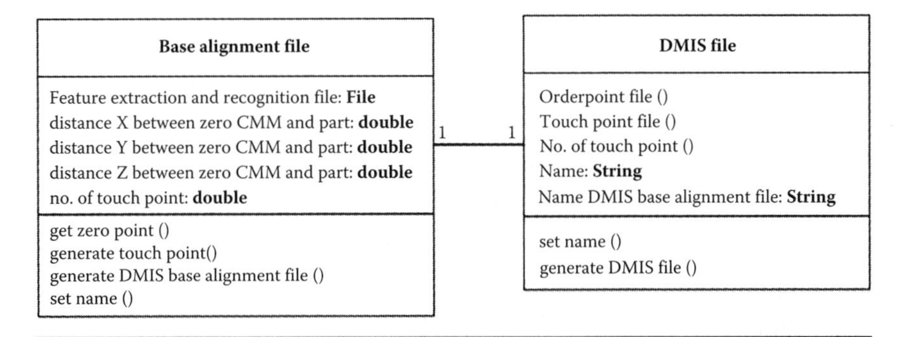

Figure 8.19 OOP class diagram of generation DMIS base-alignment and GD&T file.

file, and set name of the file. The DMIS test class consists of importing of orderpoint (inspection) file, importing of touch point file, entering number of touch point, and entering DMIS file name actions. All these actions are carried out by these methods: generate touch point, generate DMIS test file, and set name of the file.

8.4 Summary

In this chapter, the methodology for the automatic integration of CAD/CAIP system has been presented. It includes the following procedure:

- PAD using a clustering algorithm.
- Setup planning using a numerical method (ANNs) and graphical method.
- Automatic generation of inspection plan table.
- Analysis to avoid probe collision and probe holder collision.
- Automatic generation of DMIS code programming.
- Execution of CMM using DMIS code programming.

QUESTIONS

1. Define inspection plan. What is the importance of dimensional inspection planning?
2. What do you mean by the automation of the inspection system?
3. Write down the various benefits of the integrated CAD, CAIP, and CMM systems.
4. What are the three basic components of the integrated system? Explain with a schematic diagram.
5. Discuss briefly the different components of the CAIP.
6. Write down the different procedures to obtain the input of the CAIP module from AFEM.

7. Discuss briefly the following:
   Module database
   Integration between CAD and CAI
   Generation of the inspection plan for the manufactured components
8. What are the different items required for the preparation of the inspection knowledge?
9. What is the difference between the free-form geometries and the analytic geometries?
10. What is the difference between the single and the integrated geometries?
11. Differentiate between the high-level and low-level planning.
12. What do you mean by accessibility analysis?
13. Explain the difference between the probe accessibility direction (PAD) and the approach direction depth (ADD).
14. What do you mean by the setup planning?
15. Define the preferential base.
16. Explain numerical and graphical method for the setup planning.
17. What are the basic principles of the path planning?
18. What do you mean by the probe movement?
19. What are the two different types of collisions while measurement using CMM?
20. Describe the inspection plan table.
21. Discuss the coordinate measuring machine module (CMMM).
22. What are the different steps required for the generation of the DMIS file?

References

1. A. Ajmal and S. G. Zhang, The application of a knowledge based clustering algorithm as an aid to probe selection and inspection process planning. *Proceedings of Institution Mechanical Engineers: Journal of Engineering Manufacture*, 1998;212(B):299–305.

2. C. Ye, C. Fuzhi, and H. Linchang, Integration of computer aided inspection planning with CAD using STEP. *Tsinghua Science and Technology*, 1996;1(2):67–171.
3. E. A. Nasr, A. Al-Ahmari, A. Kamrani, and O. Abdulhameed, Developing an integrated system for CAD and inspection planning, *Proceedings of the 41st International Conference on Computers & Industrial Engineering*, October 23–25, 2011, pp. 660–665.
4. A. Kamrani, E. A. Nasr, A. Al-Ahmari, O. Abdulhameed, and S. Hammad Mian, Feature-based design approach for integrated CAD and computer-aided inspection planning. *International Journal of Advanced Manufacturing Technology*, 2015;76:2159–2183.
5. M. B. Adil and H. S. Ketan, Integrating design and production planning with knowledge-based inspection planning system. *The Arabian Journal for Science and Engineering*, 2005;l(30):2B.
6. M.-W. Cho, H. Lee, G.-S. Yoon, and J. Choi, A computer aided inspection planning system for on-machine measurement-part II: Local inspection planning. *KSME International Journal*, 2004;18(8):1349–1357.
7. S. N. Spitz, A. J. Spyridi, and A. A. G. Requicha, Accessibility analysis for planning of dimensional inspection with coordinate measuring machines. *IEEE Conference on Robotics and Automation*, Los Angeles, CA, August 22–25, 1999.
8. H. Lee, M.-W. Cho, G.-S. Yoon, and J.-H. Choi, A computer-aided inspection planning system for on-machine measurement—Part I: Global inspection planning. *KSME International Journal*, 2004;18:1349–1357.
9. Y. Wu, S. Liu, and G. Zhang, Improvement of coordinate measuring machine probing accessibility. *Precision Engineering*, 2004;28(1):89–94.
10. T. C. Chang, 1990, *Expert Process Planning for Manufacturing*. Reading, MA: Addison-Wesley.
11. G. Rajamohan, M. S. Shunmugam, and G. L. Samuel, Effect of probe size and measurement strategies on assessment of free-form profile deviations using coordinate measuring machine. *Measurement*, 2011;44:832–841.
12. E. A. Nasr, A. Al-Ahmari, O. Abdulhameed, and S. Hammad Mian, Set up planning for automatic generation of inspection plan, *The International Conference on Sustainable Intelligent Manufacturing*, Lisbon, Portugal, June 26–29, 2013.

13. K.-C. Fan and M. C. Leut, Intelligent planning of CAD-directed inspection for coordinate measuring machines. *Computer Integrated Manufacturing Systems*, 1998;11(1–2):43–51.
14. J. Beg and M. S. Shunmugam, An object oriented planner for inspection of prismatic parts—OOPIPP. *International Journal of Advanced Manufacturing Technology*, 2002;19:905–916.
15. R. Raghunandan and P. V. Rao, Selection of sampling points for accurate evaluation of flatness error using coordinate measuring machine. *Journal of Materials Processing Technology*, 2008;202:240–245.

Chapter 9

Application of an Integrated System for CAD and Inspection Planning

The applications of the integrated systems are presented for different cases of the prismatic parts containing more than 30 different features.

9.1 Illustrative Example 1

The component used for first case is shown in Figures 9.1 and 9.2, respectively.

9.1.1 Feature Extraction and Recognition

The extracted manufacturing features in terms of the feature identification number (ID), feature name, feature dimensions, and feature's location relative to the original coordinates of the

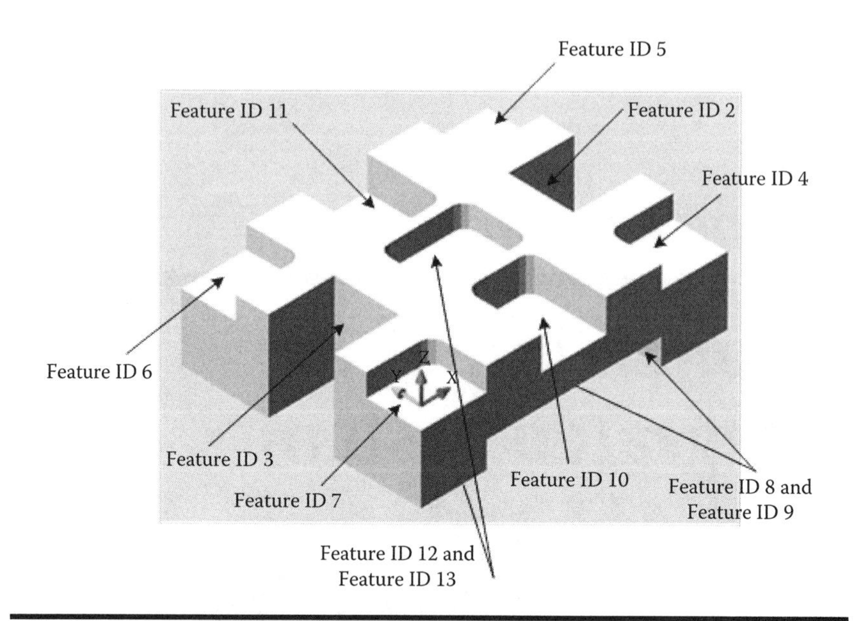

Figure 9.1 Component (solid body) with feature IDs.

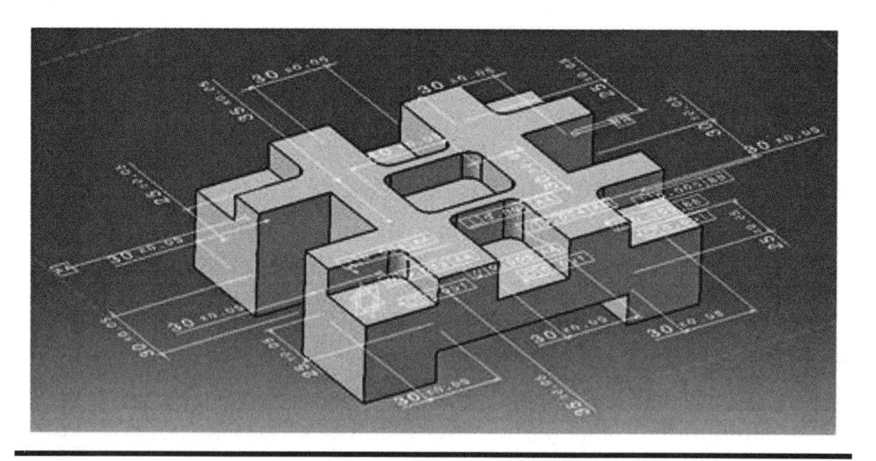

Figure 9.2 Component illustrated with datum geometrical tolerance faces.

deigned part have been shown in Table 9.1. The machining information of the designed part that includes the machining sequence, operation type, machine, cutting tool, tool/machining approach, and the removed volume for each extracted feature have been summarized in Table 9.2.

Table 9.1 Manufacturing Features and the Related Information

Feat ID	*Face ID*	*Common Edges ID*	*Location*	*Feature Name*	*Dimension*			
					L	*W*	*H*	*R*
[1]			[69] = (0,0,0)	Raw Material	140	110	40	
[2]	[2]	[158][160]	[68] = (10,0,–45)	Slot_Through	40	30	30	
[3]	[3]	[172][170]	[72] = (10,0,–15)	Slot_Through	40	30	30	
[4]	[4]	[243][13][15][247][245]	[61] = (90,0,–45)	Step_Blind_Round_Corner	10	30	25	5
[5]	[5]	[11][9][250][252][249]	[25] = (90,0,–15)	Step_Blind_Round_Corner	10	30	25	5
[6]	[6]	[253][5][7][256][255]	[55] = (0,75,–20)	Step_Blind_Round_Corner	10	30	25	5
[7]	[7]	[3][1][261][263][259]	[65] = (0,5,–20)	Step_Blind_Round_Corner	10	30	25	5
[8]	[8]	[79][77][85][75][73][90][81] [78][86][87][74][91][92]	[22] = (40,0,–15)	Slot_Blind_Round_Corner	12	80	35	5
[9]	[9]	[123][69][71][129][65][67][125] [124][70][131][130][66][127]	[38] = (0,32.5,–20)	Slot_Blind_Round_Corner	12	80	35	5
[10]	[10]	[31][29][228][27][25][231][226] [30][229][230][26][232][233]	[15] = (75,60,0)	Slot_Blind_Round_Corner	15	30	35	5

(Continued)

Table 9.1 (*Continued*) Manufacturing Features and the Related Information

Feat ID	*Face ID*	*Common Edges ID*	*Location*	*Feature Name*	*Dimension*			
					L	*W*	*H*	*R*
[11]	[11]	[236][21][23][240][17][19][238] [237][22][242][241][18][239]	[17] = (75,20.231,0)	Slot_Blind_Round_Corner	15	30	35	5
[12]	[12]	[59][57][147][55][53][150][51] [49][153][63][61][146][58] [148][149][54][151][152][50] [154][155][62][145]	[29] =(25,60.119,0)	Pocket_Blind_Round_ Corner	10	30	40	5
[13]	[13]	[43][41][217][39][37][220][35] [33][223][47][45][216][42] [218][219][38][221][222][34] [224][225][46][215]	[19] = (25,20.77,0)	Pocket_Blind_Round_ Corner	10	30	40	5

Table 9.2 Machining Information

Operating Sequence	*Feature ID*	*Feature Name*	*Operation Type*	*Machine*	*Cutting Tool*	*Tool Approach*	*Removed Volume*
[2]	[2]	Slot_Through	Slotting_ Milling	Milling	End milling cutter	[0,0,1] [0,0,–1]	36,000.000
[3]	[3]	Slot_Through	Slotting_ Milling	Milling	End milling cutter	[0,0,1] [0,0,–1]	36,000.000
[4]	[4]	Step_Blind_ Round_Corner	Shoulder_ Milling	Milling	Corner rounding milling cutter	[0,1,0] [0,0,–1] [–1,0,0]	7446.34954
[5]	[5]	Step_Blind_ Round_Corner	Shoulder_ Milling	Milling	Corner rounding milling cutter	[0,–1,0] [0,0,–1] [–1,0,0]	7446.34954
[6]	[6]	Step_Blind_ Round_Corner	Shoulder_ Milling	Milling	Corner rounding milling cutter	[0,–1,0] [0,0,–1] [1,0,0]	7446.34954
[7]	[7]	Step_Blind_ Round_Corner	Shoulder_ Milling	Milling	Corner rounding milling cutter	[0,1,0] [0,0,–1] [1,0,0]	7446.34954

(*Continued*)

Table 9.2 (*Continued*) Machining Information

Operating Sequence	*Feature ID*	*Feature Name*	*Operation Type*	*Machine*	*Cutting Tool*	*Tool Approach*	*Removed Volume*
[8]	[8]	Slot_Blind_ Round_Corner	Slotting_ Milling	Milling	Corner rounding milling cutter	[0,1,0]	33,471.23889
[9]	[9]	Slot_Blind_ Round_Corner	Slotting_ Milling	Milling	Corner rounding milling cutter	[0,–1,0]	33,471.23889
[10]	[10]	Slot_Blind_ Round_Corner	Slotting_ Milling	Milling	Corner rounding milling cutter	[0,1,0]	15,589.04861
[11]	[11]	Slot_Blind_ Round_Corner	Slotting_ Milling	Milling	Corner rounding milling cutter	[0,–1,0]	15,589.04861
[12]	[12]	Pocket_Blind_ Round_Corner	Pocket_ Milling	Milling	Corner rounding milling cutter	[0,0,1]	11,785.23889
[13]	[13]	Pocket_Blind_ Round_Corner	Pocket_ Milling	Milling	Corner rounding milling cutter	[0,0,–1]	11,785.23889

9.1.2 Inspection Plan Generation

This component has 13 features distributed on different faces, and, therefore, it cannot be inspected in a single setup. The PAD and ADD matrices have to be used for feature clustering. The determination of the probe approach direction (PAD) for every feature has been generated to represent the accessibility direction of the probe to measure all the features at every setup.

9.1.2.1 The Setup Planning of the Prismatic Parts

- First Rule (Numerical Method)
  - Geometric Extracting Entities Input: The following matrix shows the geometric entities extracted in the extraction and recognition file, and it will be used as an input in ANNs.

No. of Vertices	No. of Line Edge	No. of Circular Edge	No. of Internal Loop	No. of External Loop	No. of Concave Faces	No. of Convex Faces
176	224	80	2	92	86	6

  - PADFS Input: The following matrix shows the relationship between the feasibility to access the feature at the row setup. $F_{11} = 5$ represents the total number of features which can be accessed by $PAD_{j=1}$, at setup $S_{i=1.}$

	+x(PAD)	−x(PAD)	+y(PAD)	−y(PAD)	+z(PAD)	−z(PAD)
S(Right)	5	9	4	4	0	3
S(Left)	9	5	4	4	0	3
S(Front)	3	3	9	5	0	4
S(Rear)	3	3	5	9	0	4
S(Top)	3	3	4	4	0	5
S(Bottom)	3	3	4	4	0	9
SUM	26	26	30	30	0	28

- ANNs Output: The ANNs output has been arranged in descending order and the first order can be the right setup or the left setup ($S_1 = 25$) or ($S_2 = 25$). This setup gives all possible directions to cover all features at S_1 or S_2 setup by PAD_j. Therefore, the right is the first setup.

■ Second Rule (Graphical Method): The case contains 13 features as shown in Figure 9.3. All features can be inspected by keeping any of the possible faces ID as bases, which are arranged in ascending order of frequency: f(47, 48), f(53, 91), f34, f92, f27, or f52, as shown in Figure 9.3. Then, the best bottom face, which contains less interaction edges frequency, is face ID (47, 48), at (1, 0, 0) normal vector.

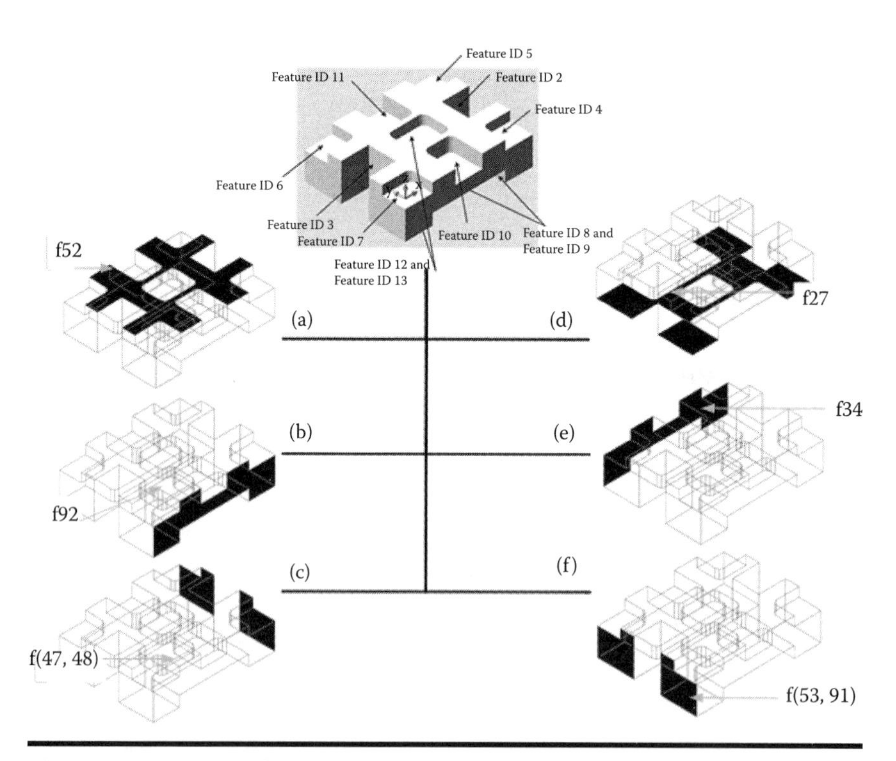

Figure 9.3 Various alternatives for the bottom (or the primary) face. Primary locating face—(a) Top (f52); (b) front (f92); (c) right (f47,48); (d) bottom (f27); (e) rear (f34); and (f) left (f53, 91).

9.1.3 Validation of the Result of the Setup Rules

This section discusses the validation between the best setup, which has been obtained from the setup rules, as compared to the worst setup orientation. The validation satisfies the orientation selected for the given part is the most suitable one because it reduces the inspection operation time.

9.1.3.1 The Best Setup

At the right setup as shown in Figure 9.4, the matrices of the PAD and ADD can be generated automatically as shown in Tables 9.3 and 9.4. The ADD matrix provides the required dimension of the probe selected such as the length of the probe, diameter of the probe sphere, and the nearest features to the clamp of the fixture in the CMM machine, which cannot be reached.

From Tables 9.3 and 9.4, the probe selected for the CMM operation is 60 mm × 3.0 mm diameter. The probe can access

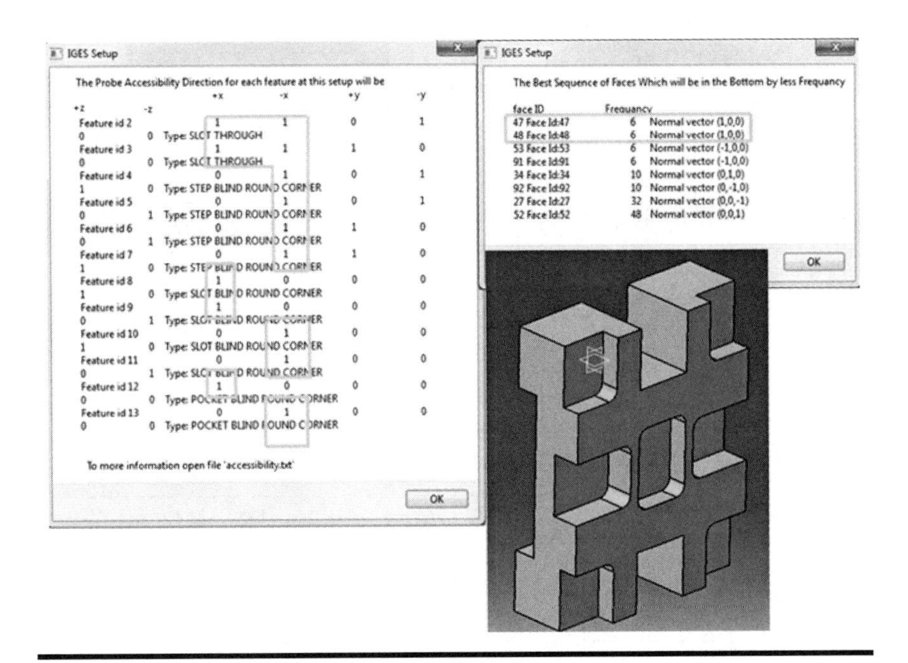

Figure 9.4 The best (right) setup and orientation.

Table 9.3 The PAD at Right Setup

	+X	−X	+Y	−Y	+Z	−Z	*Feature*
ID 2	1	1	0	1	0	0	Slot Through
ID 3	1	1	1	0	0	0	Slot Through
ID 4	0	1	0	1	1	0	Step Blind Round Corner
ID 5	0	1	0	1	0	1	Step Blind Round Corner
ID 6	0	1	1	0	0	1	Step Blind Round Corner
ID 7	0	1	1	0	1	0	Step Blind Round Corner
ID 8	1	0	0	0	1	0	Slot Blind Round Corner
ID 9	1	0	0	0	0	1	Slot Blind Round Corner
ID 10	0	1	0	0	1	0	Slot Blind Round Corner
ID 11	0	1	0	0	0	1	Slot Blind Round Corner
ID 12	1	0	0	0	0	0	Pocket Blind Round Corner
ID 13	0	1	0	0	0	0	Pocket Blind Round Corner

Table 9.4 The ADD at Right Setup

	+X	−X	+Y	−Y	+Z	−Z	*Feature*
ID 2	40	40	0	30	0	0	Slot Through
ID 3	40	40	30	0	0	0	Slot Through
ID 4	0	10	0	30	25	0	Step Blind Round Corner
ID 5	0	10	0	30	0	25	Step Blind Round Corner
ID 6	0	10	30	0	0	25	Step Blind Round Corner
ID 7	0	10	30	0	25	0	Step Blind Round Corner
ID 8	12	0	0	0	35	0	Slot Blind Round Corner
ID 9	12	0	0	0	0	35	Slot Blind Round Corner
ID 10	0	15	0	0	35	0	Slot Blind Round Corner
ID 11	0	15	0	0	0	35	Slot Blind Round Corner
ID 12	10	0	0	0	0	0	Pocket Blind Round Corner
ID 13	0	10	0	0	0	0	Pocket Blind Round Corner

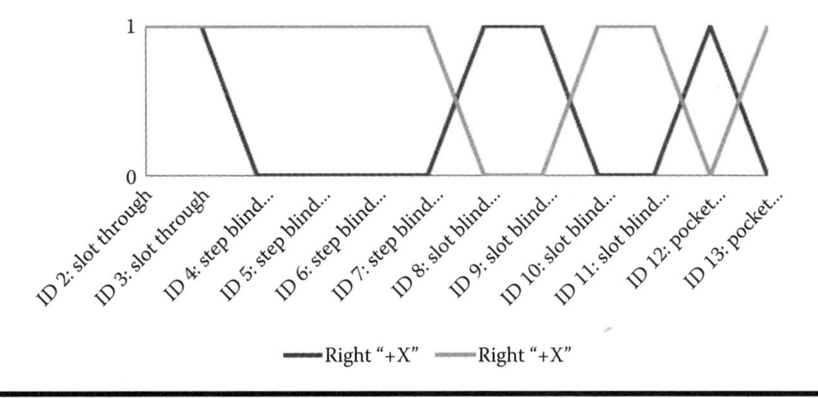

Figure 9.5 The PAD at right setup.

in (–x) direction for feature IDs (2, 3, 4, 5, 6, 7, 10, 11, and 13), (+x) direction for feature IDs (2, 3, 8, 9, and 12) as shown in Figure 9.5.

9.1.3.2 The Worst Setup

At the top setup as shown in Figure 9.6, the matrices of the PAD and ADD of the top setup are generated automatically as shown in Tables 9.5 and 9.6.

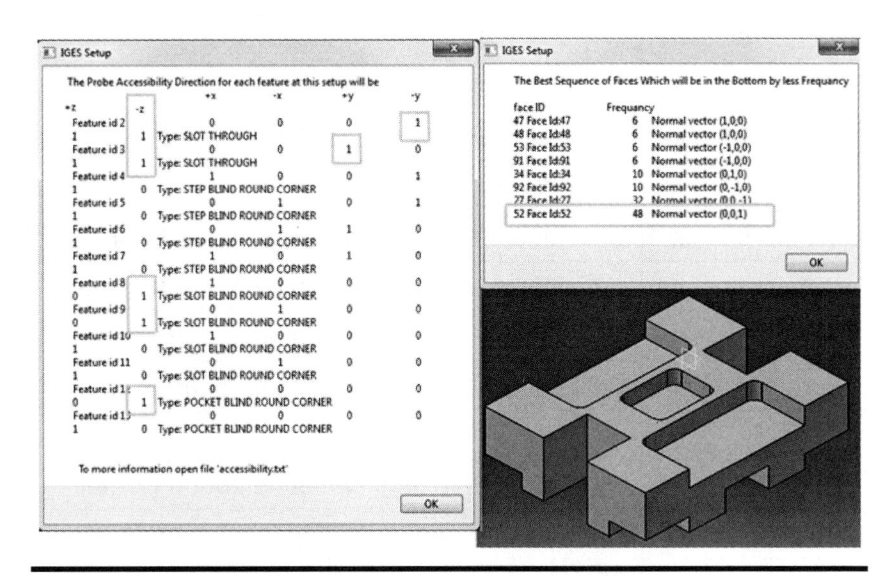

Figure 9.6 The worst (top) setup and orientation.

Table 9.5 The PAD at Top Setup

	+X	−X	+Y	−Y	+Z	−Z	
ID 2	0	0	0	1	1	1	Slot Through
ID 3	0	0	1	0	1	1	Slot Through
ID 4	1	0	0	1	1	0	Step Blind Round Corner
ID 5	0	1	0	1	1	0	Step Blind Round Corner
ID 6	0	1	1	0	1	0	Step Blind Round Corner
ID 7	1	0	1	0	1	0	Step Blind Round Corner
ID 8	1	0	0	0	0	1	Slot Blind Round Corner
ID 9	0	1	0	0	0	1	Slot Blind Round Corner
ID 10	1	0	0	0	1	0	Slot Blind Round Corner
ID 11	0	1	0	0	1	0	Slot Blind Round Corner
ID 12	0	0	0	0	0	1	Pocket Blind Round Corner
ID 13	0	0	0	0	1	0	Pocket Blind Round Corner

Table 9.6 The ADD at Top Setup

	+X	−X	+Y	−Y	+Z	−Z	
ID 2	0	0	0	30	40	40	Slot Through
ID 3	0	0	30	0	40	40	Slot Through
ID 4	25	0	0	30	10	0	Step Blind Round Corner
ID 5	0	25	0	30	10	0	Step Blind Round Corner
ID 6	0	25	30	0	10	0	Step Blind Round Corner
ID 7	25	0	30	0	10	0	Step Blind Round Corner
ID 8	35	0	0	0	12	0	Slot Blind Round Corner
ID 9	0	35	0	0	0	12	Slot Blind Round Corner
ID 10	35	0	0	0	15	0	Slot Blind Round Corner
ID 11	0	35	0	0	15	0	Slot Blind Round Corner
ID 12	0	0	0	0	0	10	Pocket Blind Round Corner
ID 13	0	0	0	0	10	0	Pocket Blind Round Corner

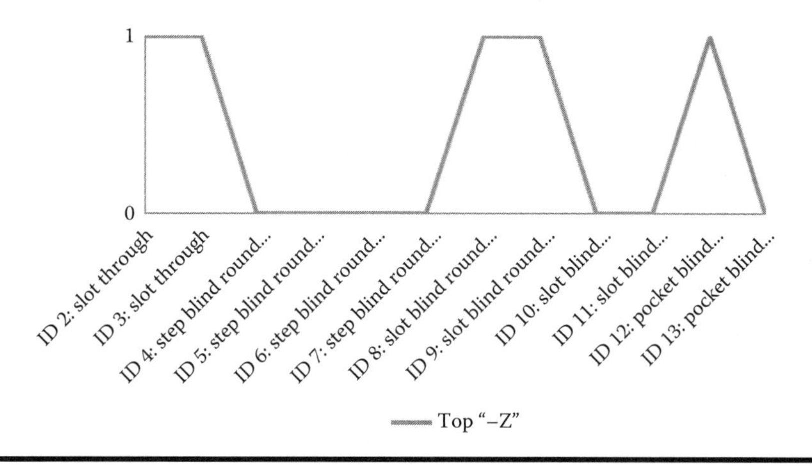

Figure 9.7 The PAD at top setup.

From Tables 9.5 and 9.6, the probe selected for CMM operation is 60 mm × 3.0 mm diameter. The probe can access in (–z) direction for feature IDs (2, 3, 8, 9, and 12). The feature IDs (4, 5, 6, 7, 10, 11, and 13) cannot be reached by the probe because of the CMM fixture system as shown in Figure 9.7.

9.1.4 Generated Inspection Table

The generated inspection plan for this component is listed in Table 9.7 for the first and second setup of the part.

9.1.5 DMIS Code Programming

Appendix A1.1 presents the contents of the DMIS code programming generated to run CMM for the third case study.

9.1.6 CMM Output

The output of the CMM inspection process is shown in Figure 9.8.

Table 9.7 Inspection Planning for the First and Second Setup of the Part

Face ID	*ID of Inspection Operation*	*Tolerance Value*	*Tool Used*	*Datum Faces ID*	*Setup of the Part*	*No. of Touch Point*	*Coordinates of Touch Point with Direction*	*Geometric Boundary of Inspection Face*
AA		—	Star probe	—	First setup	6	0,103,–33 0,77,–33 0,77,–7 0,78,–7 0,92,–17 0,96,–26	0,110,–10 0,110,–40 0,70,–40 0,70,0 0,85,0 0,85,–10
StLOne		0.05	Star probe	—	First setup	6	0,33,–33 0,7,–33 0,7,–17 0,32,–3 0,32,–7 0,26,–26	0,40,0 0,40,–40 0,0,–40 0,0,–10 0,25,–10 0,25,0
		0.05	Star probe	—	First setup	6	30,13,–3 30,13,–7 30,7,–7 30,20,–10 30,6,–14 30,4,–4	30,0,–10 30,20,–10 30,20,0 30,0,0

(*Continued*)

Table 9.7 (*Continued*) Inspection Planning for the First and Second Setup of the Part

Face ID	*ID of Inspection Operation*	*Tolerance Value*	*Tool Used*	*Datum Faces ID*	*Setup of the Part*	*No. of Touch Point*	*Coordinates of Touch Point with Direction*	*Geometric Boundary of Inspection Face*
StLTwo		0.05	Star probe	—	First setup	6	140,33,–33 140,7,–33 140,7,–17 140,32,–3 140,32,–7 140,26,–26	140,40,0 140,40,–40 140,0,–40 140,0,–10 140,25,–10 140,25,0
		0.05	Star probe	—	First setup	8	110,13,–3 110,13,–7 110,7,–7 110,20,–10 110,6,–14 110,4,–4	110,20,–10 110,20,0 110,0,0 110,0,–10

(*Continued*)

Table 9.7 (*Continued*) Inspection Planning for the First and Second Setup of the Part

Face ID	*ID of Inspection Operation*	*Tolerance Value*	*Tool Used*	*Datum Faces ID*	*Setup of the Part*	*No. of Touch Point*	*Coordinates of Touch Point with Direction*	*Geometric Boundary of Inspection Face*
StWOne		0.05	Star probe	—	First setup	10	137,0,–33 133,0,–3 103,0,–17 103,0,–21 23,0,–35 37,0,–47 117,0,–33 103,0,7 78,0,–7 78,0,–8	140,0,–40 140,0,–10 110,0,–10 110,0,–28 30,0,–28 30,0,–40 110,0,–40 110,0,0 85,0,0 85,0,–15 55,0,–15 55,0,0 30,0,0 30,0,–10 0,0,–10 0,0,–40

(*Continued*)

Table 9.7 (*Continued*) Inspection Planning for the First and Second Setup of the Part

Face ID	*ID of Inspection Operation*	*Tolerance Value*	*Tool Used*	*Datum Faces ID*	*Setup of the Part*	*No. of Touch Point*	*Coordinates of Touch Point with Direction*	*Geometric Boundary of Inspection Face*
		0.05	Star probe	—	First setup	6	18,25,–3 18,25,–7 7,25,–7 –7,25,–3 25,25,–10 11,25,–14	25,25,0 0,25,0 0,25,–10 25,25,–10
SIMLOne		0.05	Star probe	—	First setup	6	55,23,–8 55,23,–7 55,7,–7 55,16,–1 55,16,–14 55,4,–4	55,30,0 55,0,0 55,0,–15 55,30,–15
		0.05	Star probe	—	First setup	6	85,23,–8 85,23,–7 85,7,–7 85,16,–1 85,16,–14 85,4,–4	85,0,–15 85,30,–15 85,30,0 85,0,0

(*Continued*)

Table 9.7 (*Continued*) Inspection Planning for the First and Second Setup of the Part

Face ID	*ID of Inspection Operation*	*Tolerance Value*	*Tool Used*	*Datum Faces ID*	*Setup of the Part*	*No. of Touch Point*	*Coordinates of Touch Point with Direction*	*Geometric Boundary of Inspection Face*
SLLOne		0.05	Star probe	—	First setup	6	0,33,–33 0,7,–33 0,7,–17 0,32,–3 0,32,–7 0,26,–26	0,40,0 0,40,–40 0,0,–40 0,0,–10 0,25,–10 0,25,0
		0.05	Star probe	—	First setup	6	30,63,–33 30,63,–7 30,47,–7 30,56,–26 30,56,–14 30,47,–7	30,40,0 30,40,–40 30,70,–40 30,70,0
SLWOne		0.05	Star probe	—	First setup	6	23,40,–33 23,40,–7 7,40,–7 –7,40,–33 16,40,–26 16,40,–14	0,40,0 0,40,–40 30,40,–40 30,40,0

(*Continued*)

Table 9.7 (*Continued*) Inspection Planning for the First and Second Setup of the Part

Face ID	*ID of Inspection Operation*	*Tolerance Value*	*Tool Used*	*Datum Faces ID*	*Setup of the Part*	*No. of Touch Point*	*Coordinates of Touch Point with Direction*	*Geometric Boundary of Inspection Face*
		0.05	Star probe	—	First setup	6	23,70,–33 7,70,–33 7,70,–7 23,70,7 16,70,–26 14,70,–26	30,70,0 30,70,–40 0,70,–40 0,70,0
SLLTwo		0.05	Star probe	—	First setup	6	140,33,–33 140,7,–33 140,7,–17 140,32,–3 140,32,–7 140,26,–26	140,40,0 140,40,–40 140,0,–40 140,0,–10 140,25,–10 140,25,0
		0.05	Star probe	—	First setup	6	110,63,–33 110,47,–33 110,47,–7 110,56,–26 110,54,–26 110,47,–7	110,70,0 110,70,–40 110,40,–40 110,40,0

(*Continued*)

Table 9.7 (*Continued*) Inspection Planning for the First and Second Setup of the Part

Face ID	*ID of Inspection Operation*	*Tolerance Value*	*Tool Used*	*Datum Faces ID*	*Setup of the Part*	*No. of Touch Point*	*Coordinates of Touch Point with Direction*	*Geometric Boundary of Inspection Face*
SLWTwo		0.05	Star probe	—	First setup	6	133,70,–33 117,70,–33 117,70,–7 133,70,7 126,70,–26 124,70,–26	140,70,0 140,70,–40 110,70,–40 110,70,0
		0.05	Star probe	—	First setup	6	133,40,–33 133,40,–7 117,40,–7 117,40,–33 126,40,–26 126,40,–14	110,40,0 110,40,–40 140,40,–40 140,40,0
StLThre		0.05	Star probe	—	First setup	6	0,103,–33 0,77,–33 0,77,–7 0,78,–7 0,92,–17 0,96,–26	0,110,–10 0,110,–40 0,70,–40 0,70,0 0,85,0 0,85,–10

(*Continued*)

Table 9.7 (*Continued*) Inspection Planning for the First and Second Setup of the Part

Face ID	*ID of Inspection Operation*	*Tolerance Value*	*Tool Used*	*Datum Faces ID*	*Setup of the Part*	*No. of Touch Point*	*Coordinates of Touch Point with Direction*	*Geometric Boundary of Inspection Face*
		0.05	Star probe	—	First setup	6	30,103,–3 30,97,–3 30,97,–7 30,96,4 30,90,–10 30,90,0	30,90,0 30,110,0 30,110,–10 30,90,–10
SIMLTwo		0.05	Star probe	—	First setup	6	85,103,–8 85,87,–8 85,87,–7 85,96,–1 85,94,–1 85,87,–7	85,80,0 85,110,0 85,110,–15 85,80,–15
		0.05	Star probe	—	First setup	6	55,103,–8 55,87,–8 55,87,–7 55,96,–1 55,94,–1 55,87,–7	55,110,–15 55,80,–15 55,80,0 55,110,0

(*Continued*)

Table 9.7 (*Continued*) Inspection Planning for the First and Second Setup of the Part

Face ID	*ID of Inspection Operation*	*Tolerance Value*	*Tool Used*	*Datum Faces ID*	*Setup of the Part*	*No. of Touch Point*	*Coordinates of Touch Point with Direction*	*Geometric Boundary of Inspection Face*
StLFour		0.05	Star probe	—	First setup	6	140,103,–33 140,103,–17 140,78,–3 140,78,–7 140,77,–7 140,96,–26	140,110,–40 140,110,–10 140,85,–10 140,85,0 140,70,0 140,70,–40
		0.05	Star probe	—	First setup	6	110,103,–3 110,97,–3 110,97,–7 110,96,4 110,90,–10 110,97,–7	110,90,–10 110,90,0 110,110,0 110,110,–10
PockMW		0.05	Star probe	—	First setup	6	78,70,–3 78,70,–7 62,70,–7 62,70,–3 85,70,–10 71,70,–14	85,70,0 55,70,0 55,70,–10 85,70,–10

(*Continued*)

Table 9.7 (*Continued*) Inspection Planning for the First and Second Setup of the Part

Face ID	*ID of Inspection Operation*	*Tolerance Value*	*Tool Used*	*Datum Faces ID*	*Setup of the Part*	*No. of Touch Point*	*Coordinates of Touch Point with Direction*	*Geometric Boundary of Inspection Face*
		0.05	Star probe	—	First setup	6	78,40,–3 78,40,–7 62,40,–7 62,40,–3 85,40,–10 71,40,–14	55,40,–10 85,40,–10 85,40,0 55,40,0
StOneFP	⊥	0.001	Star probe	AA	First setup	6	18,25,–3 18,25,–7 7,25,–7 –7,25,–3 25,25,–10 11,25,–14	25,25,0 0,25,0 0,25,–10 25,25,–10
StOneLP	//	0.001	Star probe	AA	First setup	6	30,13,–3 30,13,–7 30,7,–7 30,20,–10 30,6,–14 30,4,–4	30,0,–10 30,20,–10 30,20,0 30,0,0

(*Continued*)

Table 9.7 (*Continued*) Inspection Planning for the First and Second Setup of the Part

Face ID	*ID of Inspection Operation*	*Tolerance Value*	*Tool Used*	*Datum Faces ID*	*Setup of the Part*	*No. of Touch Point*	*Coordinates of Touch Point with Direction*	*Geometric Boundary of Inspection Face*
StOneBF	⏥	0.001	Star probe	AA	First setup	6	73,35,–8 67,35,–8 67,35,–7 73,35,7 66,35,–1 74,35,–1	80,35,–15 60,35,–15 60,35,0 80,35,0

Measurement			Upper tolerance	Lower tolerance	Error
StLOne_0-StLOne_1-genera_0	39.6975	30.0000	0.0500	−0.0500	9.6975
StLTwo_2-StLTwo_3-genera_1	39.6865	30.0000	0.0500	−0.0500	9.6865
SlMLOne_4-SlMLOne_5-genera_2	30.0052	30.0000	0.0500	−0.0500	0.0052
SLLOne_6-SLLOne_7-genera_3	50.0563	30.0000	0.0500	−0.0500	20.0563
SLWOne_8-SLWOne_9-genera_4	30.0274	30.0000	0.0500	−0.0500	0.0274
SLLTwo_10-SLLTwo_11-genera_5	49.9298	30.0000	0.0500	−0.0500	19.9298
SLWTwo_12-SLWTwo_13-genera_6	30.0092	30.0000	0.0500	−0.0500	0.0092
StLThre_14-StLThre_15-genera_7	44.4897	30.0000	0.0500	−0.0500	14.4897
SlMLTwo_16-SlMLTwo_17-genera_8	30.0052	30.0000	0.0500	−0.0500	0.0052

Figure 9.8 Output after running the DMIS code.

9.2 Illustrative Example 2

The component for the second case is shown in Figures 9.9 and 9.10.

9.2.1 Feature Extraction and Recognition

The extracted manufacturing features in terms of feature identification number (ID), feature name, feature dimensions, and feature's location relative to the original coordinates of the deigned part are listed in Table 9.8. Machining information of the designed part that includes the machining sequence, the operation type, the machine,

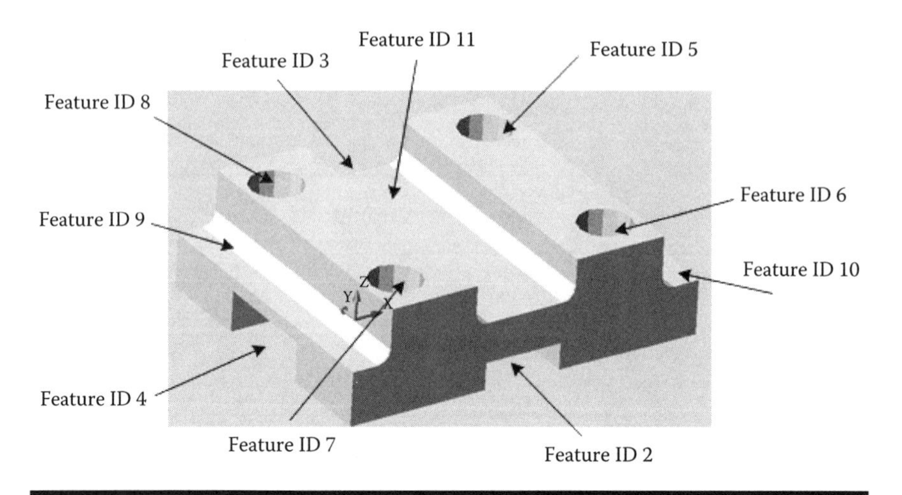

Figure 9.9 Component (solid body) with feature IDs.

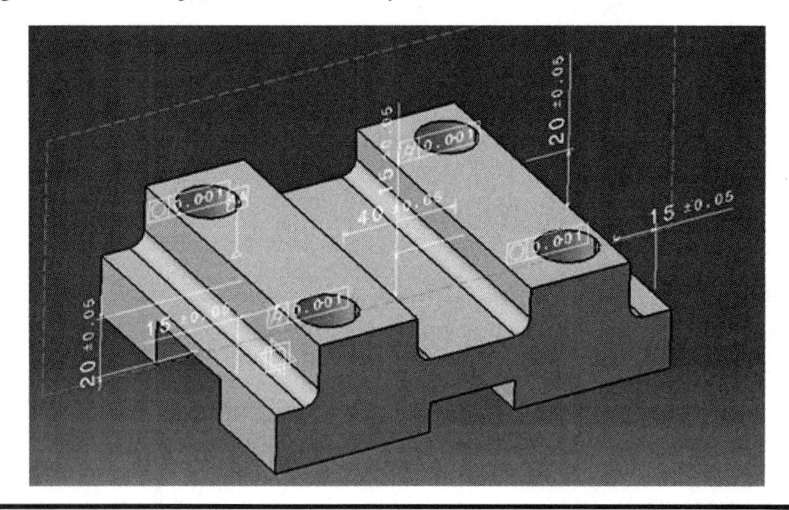

Figure 9.10 Component illustrated with datum geometrical tolerance faces.

the cutting tool, the tool/machining approach, and the removed volume for each extracted feature are summarized in Table 9.9.

9.2.2 Inspection Plan Generation

This selected component consists of 13 features distributed on different faces, and therefore it cannot be inspected in a single

Table 9.8 Manufacturing Features and the Related Information

Feat ID	*Face ID*	*Common Edges ID*	*Location*	*Feature Name*	*Dimension*			
					L	*W*	*H*	*R*
[1]			[69] = (0,0,0)	Raw Material	140	110	40	
[2]	[2]	[85][92]	[47] = (55,0,–10)	Slot Through	35	30	10	
[3]	[3]	[90][95]	[66] = (55,75,–10)	Slot Through	35	30	10	
[4]	[4]	[108][114]	[80] = (0,45,–15)	Slot Through	140	40	15	
[5]	[5]	[1][2][3][4][83][67]	(112.5,92.5,0)	Hole Through			40	10
[6]	[6]	[5][6][7][8][82][66]	(112.5,17.5,0)	Hole Through			40	10
[7]	[7]	[9][10][11][12][81][75]	(27.5,17.5,0)	Hole Through			40	10
[8]	[8]	[29][30][31][32][33][34]	(27.5,92.5,0)	Hole Through			40	10
[9]	[9]	[69][17][19]	[17] = (13,0,0)	Step Through Round Corner	110	20	10	
[10]	[10]	[76][13][15]	[14] = (126.5,0,–21.5)	Step Through Round Corner	110	20	10	
[11]	[11]	[27][25][36][42][21][23]	[25] = (56.5,0,–21.5)	Slot Through Round Corner	110	30	20	

Table 9.9 Machining Information

Operating Sequence	*Feature ID*	*Feature Name*	*Operation Type*	*Machine*	*Cutting Tool*	*Tool Approach*	*Removed Volume*
[2]	[2]	Slot Through	Slotting Milling	Milling	End milling cutter	[0,1,0] [0,–1,0]	21,000.00
[3]	[3]	Slot Through	Slotting Milling	Milling	End milling cutter	[0,1,0] [0,–1,0]	21,000.00
[4]	[4]	Slot Through	Slotting Milling	Milling	End milling cutter	[1,0,0] [–1,0,0]	84,000.00
[5]	[5]	Hole Through	Drilling	Drilling	Twist drill	[0,0,1] [0,0,–1]	12,566.37040
[6]	[6]	Hole Through	Drilling	Drilling	Twist drill	[0,0,1] [0,0,–1]	12,566.37040
[7]	[7]	Hole Through	Drilling	Drilling	Twist drill	[0,0,1] [0,0,–1]	12,566.37040

(*Continued*)

Table 9.9 (*Continued*) Machining Information

Operating Sequence	*Feature ID*	*Feature Name*	*Operation Type*	*Machine*	*Cutting Tool*	*Tool Approach*	*Removed Volume*
[8]	[8]	Hole Through	Drilling	Drilling	Twist drill	[0,0,1] [0,0,–1]	12,566.37040
[9]	[9]	Step Through Round Corner	Shoulder Milling	Milling	Corner rounding milling cutter	[0,1,0] [0,–1,0]	21,409.84491
[10]	[10]	Step Through Round Corner	Shoulder Milling	Milling	Corner rounding milling cutter	[0,1,0] [0,–1,0]	21,409.84491
[11]	[11]	Slot Through Round Corner	Slotting Milling	Milling	Corner rounding milling cutter	[0,1,0] [0,–1,0]	64,819.68983

setup. The PAD and ADD matrices have been used for feature clustering.

9.2.2.1 The Setup Planning of the Prismatic Parts

- First Rule (Numerical Method)
  - Geometric Extracting Entities Input: The following matrix shows the geometric entities in the extraction and recognition file, which can be used as an input in ANNs.

No. of Vertices	No. of Line Edge	No. of Circular Edge	No. of Internal Loop	No. of External Loop	No. of Concave Faces	No. of Convex Faces
80	88	32	8	42	27	15

  - PADFS Input: The following matrix shows the relationship between the feasibility to access the feature at the row setup. $F_{11} = 7$ represents the total number of features which can be accessed by $PAD_{j=1}$, at setup $S_{i=1.}$

	+x(PAD)	−x(PAD)	+y(PAD)	−y(PAD)	+z(PAD)	−z(PAD)
S(Right)	7	7	5	5	0	2
S(Left)	7	7	5	5	0	2
S(Front)	2	2	7	7	0	5
S(Rear)	2	2	7	7	0	5
S(Top)	2	2	5	5	0	7
S(Bottom)	2	2	5	5	0	7
SUM	22	22	34	34	0	28

  - ANNs Output: The ANNs output has been arranged in descending order and the first order can be the right setup or the left setup ($S_1 = 26$) or ($S_2 = 26$). This setup gives all the possible directions to cover all the features at S_1 or S_2 setup by PAD_j. Therefore, the right is the first setup.

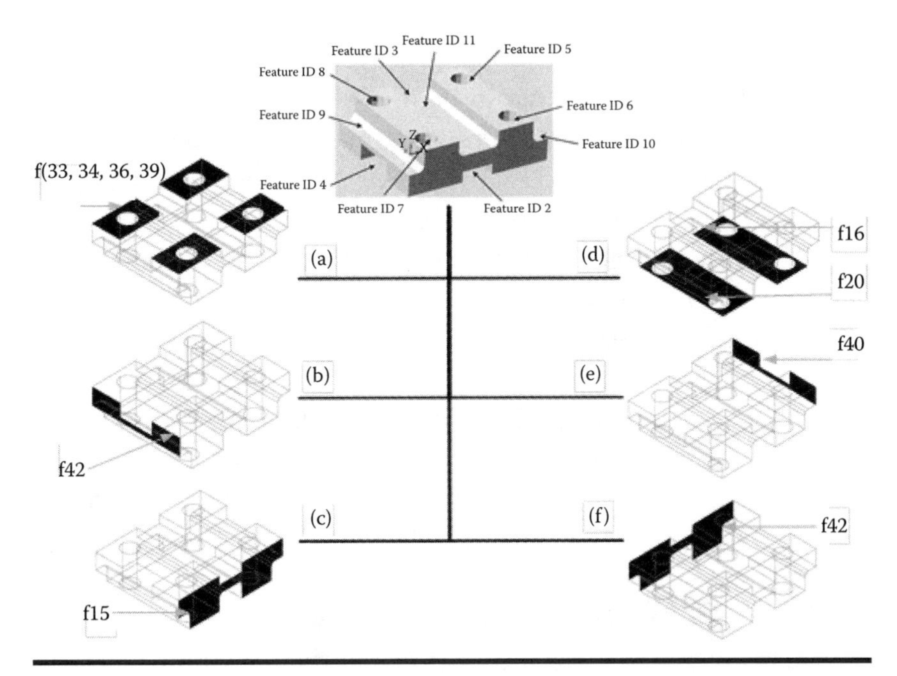

Figure 9.11 Various alternatives for the bottom (or the primary) face. Primary locating face—(a) bottom (f33,34,36,39); (b) right (f42); (c) rear (f15); (d) top (f16, f20); (e) left (f40); and (f) front (f42).

- Second Rule (Graphical Method): This case contains 11 features as shown in Figure 9.9 and all the features can be inspected by keeping any of the possible faces ID as bases, which are arranged in ascending order of frequency: f40, f41, f15, f42, f(16, 20), or f(33, 34, 36, 39), as shown in Figure 9.11. The ideal bottom face, which contains less interaction edges frequency, is face ID 40 or ID 41 at (–1,0,0) or (1,0,0) normal vector.

9.2.2.2 Validation of the Result of the Setup Rules

This section describes the comparison between the best possible setup orientations and the worst setup orientation. This validation satisfies the orientation selection to avoid increase in inspection time.

9.2.2.2.1 The Best Setup

At the right setup as shown in Figure 9.12, the matrices of the PAD and ADD of the right setup is generated automatically as shown in Tables 9.10 and 9.11. The ADD matrix give us the required dimension for probe selection such as the length of the probe, the diameter of the probe sphere, and the nearest features to the clamp of the fixture in the CMM machine, which cannot be reached.

From Tables 9.8 and 9.9, the type of the probe selected for the CMM operation is the probe with dimensions 60 mm × 3.0 mm diameter. The probe can access in (–x) direction for feature IDs (2, 3, 4, 5, 6, 7, and 8), (+x) direction for feature IDs (5, 6, 7, 8, 9, 10, and 11) as shown in Figure 9.13.

9.2.2.2.2 The Worst Setup

As shown in Figure 9.14, the matrices of the PAD and ADD of the top setup are generated automatically as shown in Tables 9.12 and 9.13.

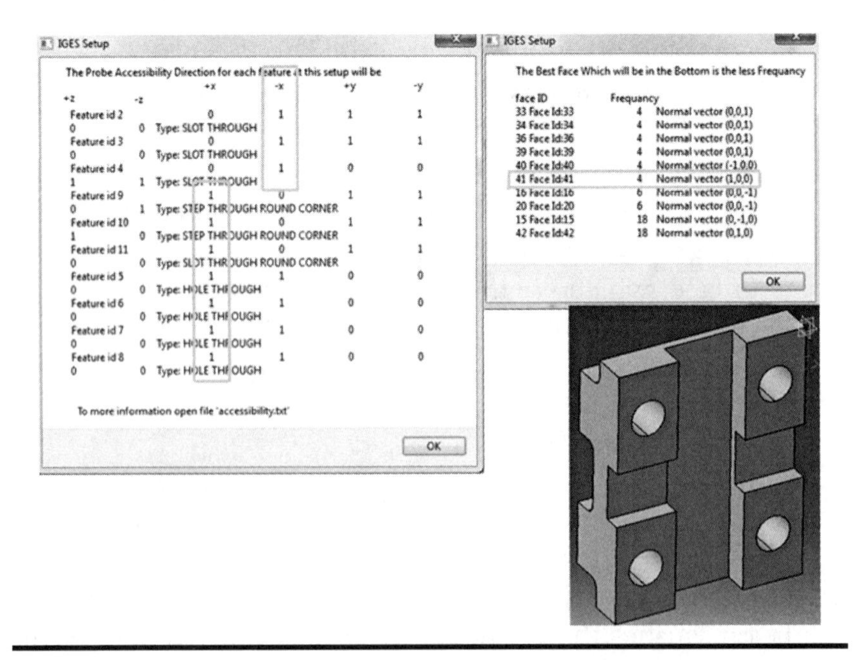

Figure 9.12 The best (right) setup and orientation.

Table 9.10 The PAD at Right Setup

PAD						
+X	–X	+Y	–Y	+Z	–Z	
0	1	1	1	0	0	ID 2: Slot Through
0	1	1	1	0	0	ID 3: Slot Through
0	1	0	0	1	1	ID 4: Slot Through
1	0	1	1	0	1	ID 9: Step Through Round Corner
1	0	1	1	1	0	ID 10: Step Through Round Corner
1	0	1	1	0	0	ID 11: Slot Through Round Corner
1	1	0	0	0	0	ID 5: Hole Through
1	1	0	0	0	0	ID 6: Hole Through
1	1	0	0	0	0	ID 7: Hole Through
1	1	0	0	0	0	ID 8: Hole Through

Table 9.11 The ADD at Right Setup

PAD						
+X	–X	+Y	–Y	+Z	–Z	
0	10	35	35	0	0	ID 2: Slot Through
0	10	35	35	0	0	ID 3: Slot Through
0	15	0	0	140	140	ID 4: Slot Through
10	0	110	110	0	20	ID 9: Step Through Round Corner
10	0	110	110	20	0	ID 10: Step Through Round Corner
25	0	110	110	0	0	ID 11: Slot Through Round Corner
40	40	0	0	0	0	ID 5: Hole Through
40	40	0	0	0	0	ID 6: Hole Through
40	40	0	0	0	0	ID 7: Hole Through
40	40	0	0	0	0	ID 8: Hole Through

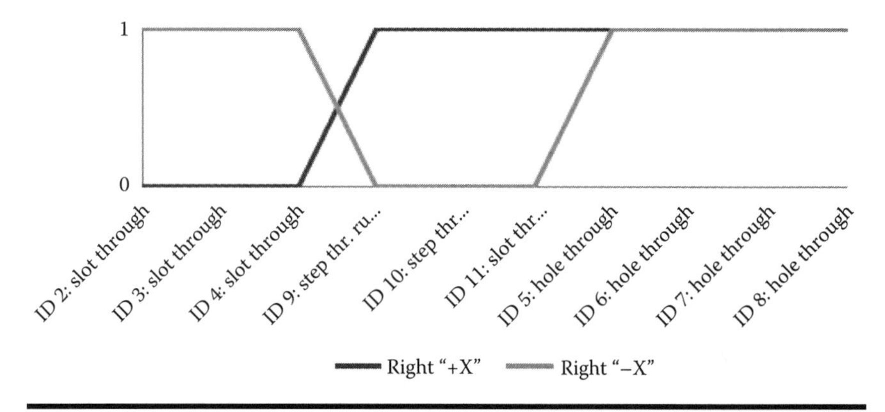

Figure 9.13 The PAD at right setup.

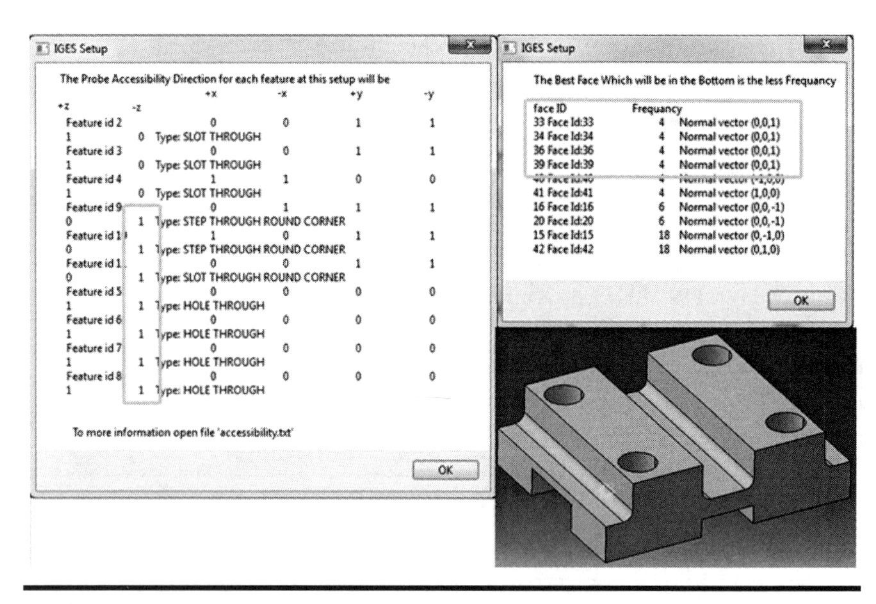

Figure 9.14 The worst (top) setup and orientation.

From Tables 9.12 and 9.13, the probe selected for the CMM operation is the probe with dimensions 60 mm × 3.0 mm diameter. The probe can access in (−z) direction for feature IDs (5, 6, 7, 8, 9, 10, and 11). The feature IDs (2, 3, and 4) cannot be reached by the probe because the CMM fixture system as shown in Figure 9.15.

Table 9.12 The PAD at Top Setup

PAD						
+X	*−X*	*+Y*	*−Y*	*+Z*	*−Z*	
0	0	1	1	1	0	ID 2: Slot Through
0	0	1	1	1	0	ID 3: Slot Through
1	1	0	0	1	0	ID 4: Slot Through
0	1	1	1	0	1	ID 9: Step Through Round Corner
1	0	1	1	0	1	ID 10: Step Through Round Corner
0	0	1	1	0	1	ID 11: Slot Through Round Corner
0	0	0	0	1	1	ID 5: Hole Through
0	0	0	0	1	1	ID 6: Hole Through
0	0	0	0	1	1	ID 7: Hole Through
0	0	0	0	1	1	ID 8: Hole Through

Table 9.13 The ADD at Top Setup

PAD						
+X	−X	+Y	−Y	+Z	−Z	
0	0	35	35	10	0	ID 2: Slot Through
0	0	35	35	10	0	ID 3: Slot Through
140	140	0	0	15	0	ID 4: Slot Through
0	10	110	110	0	20	ID 9: Step Through Round Corner
10	0	110	110	0	20	ID 10: Step Through Round Corner
0	0	110	110	0	25	ID 11: Slot Through Round Corner
0	0	0	0	40	40	ID 5: Hole Through
0	0	0	0	40	40	ID 6: Hole Through
0	0	0	0	40	40	ID 7: Hole Through
0	0	0	0	40	40	ID 8: Hole Through

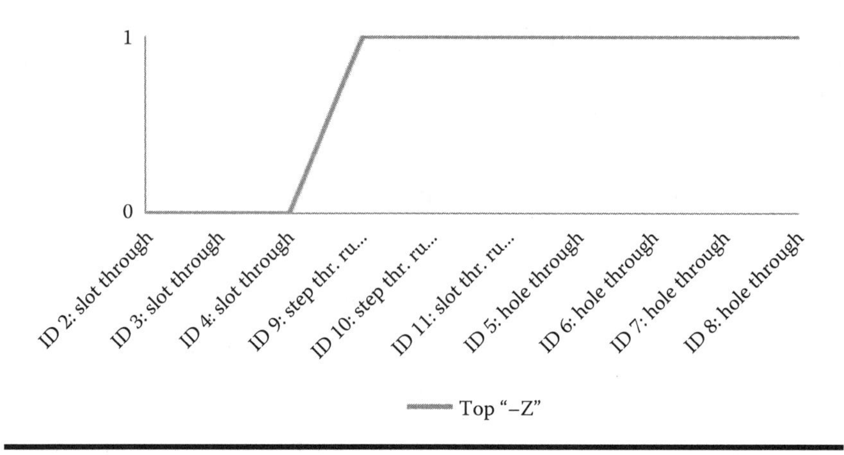

Figure 9.15 The PAD at top setup.

9.2.3 Generated Inspection Table

The generated inspection plan for the component is listed in Table 9.14 for the first and second setup of the part.

9.2.4 DMIS Code Programming

Appendix A1.2 presents the contents of the DMIS code programming generated to run CMM for the first case study.

9.2.5 CMM Output

The output of the CMM inspection process is shown in Figure 9.16.

9.3 Illustrative Example 3

The proposed methodology is used for the component shown in Figures 9.17 and 9.18 has been explained in the following sections.

Table 9.14 Inspection Planning for the First and Second Setup of the Part

Face ID	ID of Inspection Operation	Tolerance Value	Tool Used	Datum Faces ID	Setup of the Part	No. of Touch Point	Coordinates of Touch Point with Direction	Geometric Boundary of Inspection Face
AA		—	Star probe	—	First setup	6	43,7,0 43,103,0 22,103,0 36,14,0 36,96,0 22,103,0	50,110,0 15,110,0 15,0,0 50,0,0
StOneL		0.05	Star probe	—	First setup	6	0,103,–33 0,82,–33 0,68,–18 0,28,–32 0,28,–33 0,7,–33	0,110,–20 0,110,–40 0,75,–40 0,75,–25 0,35,–25 0,35,–40 0,0,–40 0,0,–20

(*Continued*)

Table 9.14 (*Continued*) Inspection Planning for the First and Second Setup of the Part

Face ID	*ID of Inspection Operation*	*Tolerance Value*	*Tool Used*	*Datum Faces ID*	*Setup of the Part*	*No. of Touch Point*	*Coordinates of Touch Point with Direction*	*Geometric Boundary of Inspection Face*
		0.05	Star probe	—	First setup	6	15,103,–8 15,103,–7 15,7,–7 15,96,–1 15,96,–14 15,7,–7	15,0,0 15,0,–15 15,110,–15 15,110,0
StOneDp		0.05	Star probe	—	First setup	6	3,7,–20 3,103,–20 7,103,–20 7,7,–20 –4,14,–20 10,110,–20	10,110,–20 0,110,–20 0,0,–20 10,0,–20
		0.05	Star probe	—	First setup	8	43,7,0 43,103,0 22,103,0 8,–7,0 36,14,0 36,96,0	50,110,0 15,110,0 15,0,0 50,0,0

(*Continued*)

Table 9.14 (*Continued*) Inspection Planning for the First and Second Setup of the Part

Face ID	*ID of Inspection Operation*	*Tolerance Value*	*Tool Used*	*Datum Faces ID*	*Setup of the Part*	*No. of Touch Point*	*Coordinates of Touch Point with Direction*	*Geometric Boundary of Inspection Face*
StMiddL		0.05	Star probe	—	First setup	6	50,103,–3 50,7,–3 50,7,–7 50,96,4 50,0,–10 50,7,–7	50,110,0 50,110,–10 50,0,–10 50,0,0
		0.05	Star probe	—	First setup	8	90,103,–3 90,103,–7 90,7,–7 90,110,–10 90,96,–14 90,7,–7	90,0,0 90,0,–10 90,110,–10 90,110,0
StMidDp		0.05	Star probe	—	First setup	6	118,7,0 118,103,0 97,103,0 83,–7,0 111,14,0 111,96,0	90,0,0 125,0,0 125,110,0 90,110,0

(*Continued*)

Table 9.14 (*Continued*) Inspection Planning for the First and Second Setup of the Part

Face ID	*ID of Inspection Operation*	*Tolerance Value*	*Tool Used*	*Datum Faces ID*	*Setup of the Part*	*No. of Touch Point*	*Coordinates of Touch Point with Direction*	*Geometric Boundary of Inspection Face*
		0.05	Star probe	—	First setup	6	78,7,–15 78,103,–15 62,103,–15 48,–7,–15 71,14,–15 71,96,–15	55,0,–15 85,0,–15 85,110,–15 55,110,–15
StTwoDp		0.05	Star probe	—	First setup	6	133,7,–20 133,103,–20 137,103,–20 123,–7,–20 126,14,–20 136,110,–20	130,0,–20 140,0,–20 140,110,–20 130,110,–20
		0.05	Star probe	—	First setup	6	118,7,0 118,103,0 97,103,0 83,–7,0 111,14,0 111,96,0	90,0,0 125,0,0 125,110,0 90,110,0

(*Continued*)

Table 9.14 (*Continued*) Inspection Planning for the First and Second Setup of the Part

Face ID	*ID of Inspection Operation*	*Tolerance Value*	*Tool Used*	*Datum Faces ID*	*Setup of the Part*	*No. of Touch Point*	*Coordinates of Touch Point with Direction*	*Geometric Boundary of Inspection Face*
StTwoL		0.05	Star probe	—	First setup	6	140,103,–33 140,103,–27 140,7,–27 140,7,–33 140,68,–33 140,68,–32	140,110,–40 140,110,–20 140,0,–20 140,0,–40 140,75,–40 140,75,–25 140,35,–25 140,35,–40
		0.05	Star probe	—	First setup	6	125,103,–8 125,103,–7 125,7,–7 125,96,–1 125,96,–14 125,7,–7	125,0,–15 125,110,–15 125,110,0 125,0,0

(*Continued*)

Table 9.14 (*Continued*) Inspection Planning for the First and Second Setup of the Part

Face ID	*ID of Inspection Operation*	*Tolerance Value*	*Tool Used*	*Datum Faces ID*	*Setup of the Part*	*No. of Touch Point*	*Coordinates of Touch Point with Direction*	*Geometric Boundary of Inspection Face*
First Hole	⌭	0.001	Star probe	—	First setup	8	37.5,17.5,–4,1,0,0 27.5,27.5,–4,0,–1,0 17.5,17.5,–4,–1,0,0 27.5,7.5,–4,0,1,0 37.5,17.5,–8,1,0,0 27.5,27.5,–8,0,–1,0 17.5,17.5,–8,–1,0,0 27.5,7.5,–8,0,1,0	Center point 27.5,17.5,–20 — Radius — Height 40
Second Hole	○	0.001	Star probe	—	First setup	8	122.5,17.5,–4,1,0,0 112.5,27.5,–4,0,-1,0 102.5,17.5,–4,–1,0,0 112.5,7.5,–4,0,1,0 122.5,17.5,–8,1,0,0 112.5,27.5,–8,0,–1,0 102.5,17.5,–8,–1,0,0 112.5,7.5,–8,0,1,0	Center 112.5,17.5,–20 — Radius 10 — Height 40

(*Continued*)

Table 9.14 (*Continued*) Inspection Planning for the First and Second Setup of the Part

Face ID	*ID of Inspection Operation*	*Tolerance Value*	*Tool Used*	*Datum Faces ID*	*Setup of the Part*	*No. of Touch Point*	*Coordinates of Touch Point with Direction*	*Geometric Boundary of Inspection Face*
Third Hole	○	0.001	Star probe	—	First setup	8	37.5,92.5,–4,1,0,0 27.5,102.5,–4,0,–1,0 17.5,92.5,–4,–1,0,0 27.5,82.5,–4,0,1,0 37.5,92.5,–8,1,0,0 27.5,102.5,–8,0,–1,0 17.5,92.5,–8,–1,0,0 27.5,82.5,–8,0,1,0	Center 27.5,92.5,–20 — Radius 10 — Height 40
Fourth Hole	⌭	0.001	Star probe	—	First setup	8	122.5,92.5,–4,1,0,0 112.5,102.5,–4,0,–1,0 102.5,92.5,–4,–1,0,0 112.5,82.5,–4,0,1,0 122.5,92.5,–8,1,0,0 112.5,102.5,–8,0,–1,0 102.5,92.5,–8,–1,0,0 112.5,82.5,–8,0,1,0	Center point 112.5,92.5,–20 — Radius 10 — Height 40

(*Continued*)

Table 9.14 (*Continued*) Inspection Planning for the First and Second Setup of the Part

Face ID	*ID of Inspection Operation*	*Tolerance Value*	*Tool Used*	*Datum Faces ID*	*Setup of the Part*	*No. of Touch Point*	*Coordinates of Touch Point with Direction*	*Geometric Boundary of Inspection Face*
Middle Step	⏥	0.001	Star probe	AA	First setup	6	118,7,0 118,103,0 97,103,0 83,–7,0 111,14,0 111,96,0	90,0,0 125,0,0 125,110,0 90,110,0

Measurement			Upper tolerance	Lower tolerance	Error
	StOneL_1-StOneL_2-genera_1				
	73.0432	15.0000	0.0500	0.0500	58.0432
	StOneDp_3-StOneDp_4-genera_3				
	24.9631	20.0000	0.0500	0.0500	4.9631
	StMiddL_5-StMiddL_6-genera_5				
	29.9159	30.0000	0.0500	0.0500	–0.0841
	StMidDp_7-StMidDp_8-genera_7				
	31.9815	20.0000	0.0500	0.0500	11.9815
	StTwoDp_9-StTwoDp_10-genera_9				
	49.2359	20.0000	0.0500	0.0500	29.2359
	StTwoL_11-StTwoL_12-genera_11				
	20.4767	15.0000	0.0500	0.0500	5.4767
	StMidRg_13-flatne_13				
	0.0000	0.0000	0.0500		0.0000
	StMidLf_15-perpen_15				
	0.0012	0.0000	0.0010		0.0012
	FrHole_16-cylind_16				
	0.0276	0.0000	0.0010		0.0276

Figure 9.16 Output after running the DMIS code.

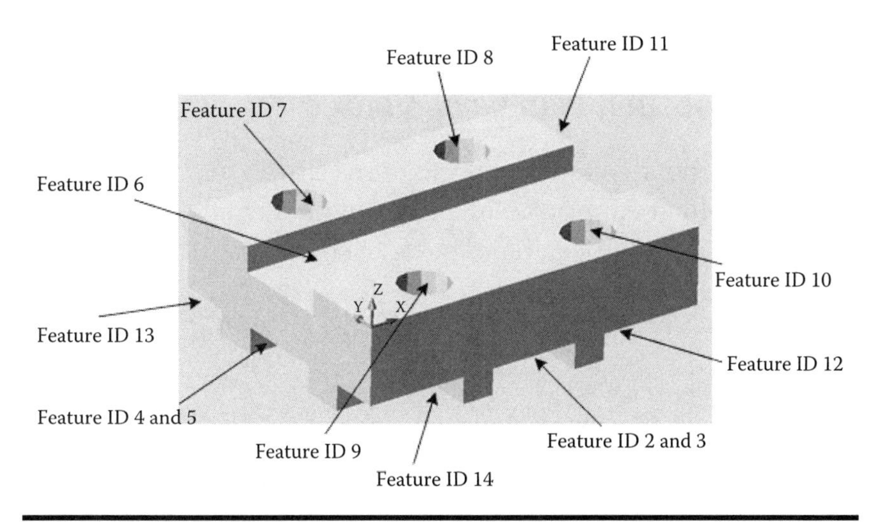

Figure 9.17 Component (solid body) with feature IDs.

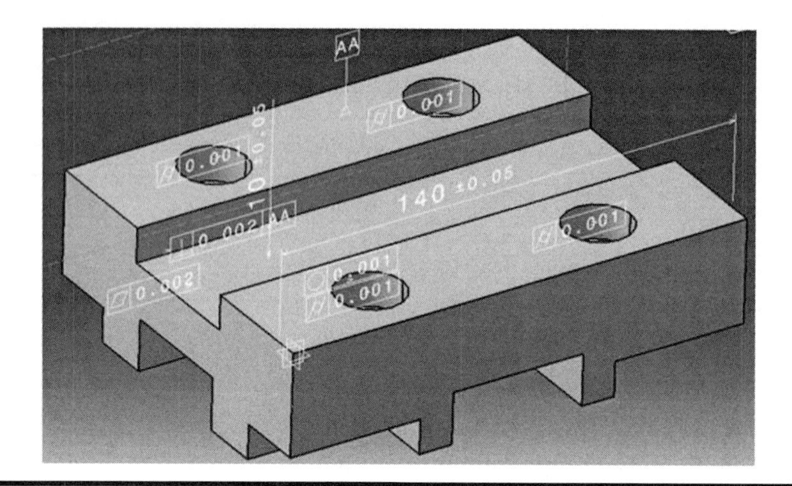

Figure 9.18 Component illustrated with datum geometrical tolerance faces.

9.3.1 Feature Extraction and Recognition

The extracted manufacturing features in terms of feature identification number (ID), feature name, feature dimensions, and features' location relative to the original coordinates of the deigned part are listed in Table 9.15. The machining information includes the machining sequence, the operation type, the machine, the cutting tool, the tool/machining approach, and the removed volume for each extracted feature are summarized in Table 9.16.

9.3.2 Inspection Plan Generation

This component consists of 13 features distributed on different faces which cannot be inspected with a single setup. The PAD and ADD matrices have been used for feature clustering.

9.3.2.1 The Setup Planning of the Prismatic Parts

- First Rule (Numerical Method)
  - Geometric Extracting Entities Input: The following matrix shows the geometric entities extracting in the extraction and recognition file, which can be used as an input in ANNs.

Table 9.15 Manufacturing Features and the Related Information

Feat ID	*Face ID*	*Common Edges ID*	*Location*	*Feature Name*	*Dimension*			
					L	*W*	*H*	*R*
[1]			[79] = (0,0,0)	Raw Material	140	110	40	
[2]	[2]	[64][70]	[55] = (52.5,0,–30)	Slot Through	37.5	72.5	10	
[3]	[3]	[74][76]	[64] = (87.5,0,–30)	Slot Through	37.5	35.5	10	
[4]	[4]	[88][89]	[74] = (0,72.5,–30)	Slot Through	52.5	35.5	10	
[5]	[5]	[103][99]	[80] = (0,37.5,–30)	Slot Through	52.5	85.5	10	
[6]	[6]	[140][142]	[93] = (0,35.5,–10)	Slot Through	140	40	10	
[7]	[7]	[1][2][3][4][138][139]	(35,92.5,0)	Hole Blind			10	10
[8]	[8]	[5][6][7][8][136][137]	(105,92.5,0)	Hole Blind			10	10
[9]	[9]	[9][10][11][12][134][135]	(35,17.5,0)	Hole Blind			10	10
[10]	[10]	[13][14][15][16][132][133]	(105,17.5,0)	Hole Blind			10	10
[11]	[11]	[31][29][34][39][36]	[30] = (101.5,91.5,–30)	Step Blind Round Corner	10	40	20	
[12]	[12]	[43][25][27][108][47]	[25] = (101.5,18.5,–30)	Step Blind Round Corner	10	20	40	
[13]	[13]	[49][21][23][110][53]	[21] = (38.5,91.5,–30)	Step Blind Round Corner	10	20	40	
[14]	[14]	[19][17][55][60][57]	[18] = (38.5,18.5,–30)	Step Blind Round Corner	10	40	20	

Table 9.16 Machining Information

Operating Sequence	*Feature ID*	*Feature Name*	*Operation Type*	*Machine*	*Cutting Tool*	*Tool Approach*	*Removed Volume*
[2]	[2]	Slot Through	Slotting milling	Milling	End milling cutter	[0,1,0] [0,–1,0]	27,187.50
[3]	[3]	Slot Through	Slotting milling	Milling	End milling cutter	[0,1,0] [0,–1,0]	13,125.00
[4]	[4]	Slot Through	Slotting milling	Milling	End milling cutter	[1,0,0] [–1,0,0]	18,375.00
[5]	[5]	Slot Through	Slotting milling	Milling	End milling cutter	[1,0,0] [–1,0,0]	45,937.50
[6]	[6]	Slot Through	Slotting milling	Milling	End milling cutter	[1,0,0] [–1,0,0]	56,000.00
[7]	[7]	Hole Blind	Counter boring drilling	Drilling	Drill + counter bore drill	[0,0,–1]	6283.18520
[8]	[8]	Hole Blind	Counter boring drilling	Drilling	Drill + counter bore drill	[0,0,–1]	6283.18520
[9]	[9]	Hole Blind	Counter boring Drilling	Drilling	Drill + counter bore drill	[0,0,–1]	6283.18520

(Continued)

Table 9.16 (*Continued*) Machining Information

Operating Sequence	*Feature ID*	*Feature Name*	*Operation Type*	*Machine*	*Cutting Tool*	*Tool Approach*	*Removed Volume*
[10]	[10]	Hole Blind	Counter boring drilling	Drilling	Drill + counter bore drill	[0,0,–1]	6283.18520
[11]	[11]	Step Blind Round Corner	Shoulder milling	Milling	Corner rounding milling cutter	[0,–1,0] [0,0,1] [–1,0,0]	7946.34954
[12]	[12]	Step Blind Round Corner	Shoulder milling	Milling	Corner rounding milling cutter	[–1,0,0] [0,0,1] [0,1,0]	7946.34954
[13]	[13]	Step Blind Round Corner	Shoulder milling	Milling	Corner rounding milling cutter	[1,0,0] [0,0,1] [0,–1,0]	7946.34954
[14]	[14]	Step Blind Round Corner	Shoulder milling	Milling	Corner rounding milling cutter	[0,1,0] [0,0,1] [1,0,0]	7946.34954

No. of Vertices	No. of Line Edge	No. of Circular Edge	No. of Internal Loop	No. of External Loop	No. of Concave Faces	No. of Convex Faces
96	112	32	4	54	48	6

– PADFS Input: The following matrix shows the relationship between the feasibility to access the feature at the row setup. $F_{11} = 8$ and represents the total number of features which can be accessed by $PAD_{j=1}$, at setup $S_{i=1}$.

	+x(PAD)	−x(PAD)	+y(PAD)	−y(PAD)	+z(PAD)	−z(PAD)
S(Right)	8	3	6	6	0	7
S(Left)	3	8	6	6	0	7
S(Front)	6	6	3	8	0	7
S(Rear)	6	6	8	3	0	7
S(Top)	7	7	6	6	0	8
S(Bottom)	7	7	6	6	0	3
SUM	37	37	35	35	0	39

– ANNs' Output: The ANNs' output has been arranged in the descending order. The first order is the top setup followed by the front setup as the next set up ($S_1 = 34$) or ($S_2 = 30$). This is all the possible directions inspect features at S_1 or S_2 setup by PAD_j. Therefore, the front is the first setup.

- Second Rule (Graphical Method): The case consists of 14 features as shown in Figure 9.19 which can be inspected by selecting any of the possible faces ID as bases, which are arranged in an ascending order of frequency: f27, f54, f(52, 53), f38, f40 or f(23, 24, 29, 39) as shown in Figure 9.19. Then, the best bottom face, which contains less interaction edges frequency, is face ID 27 at (0, –1, 0) normal vector.

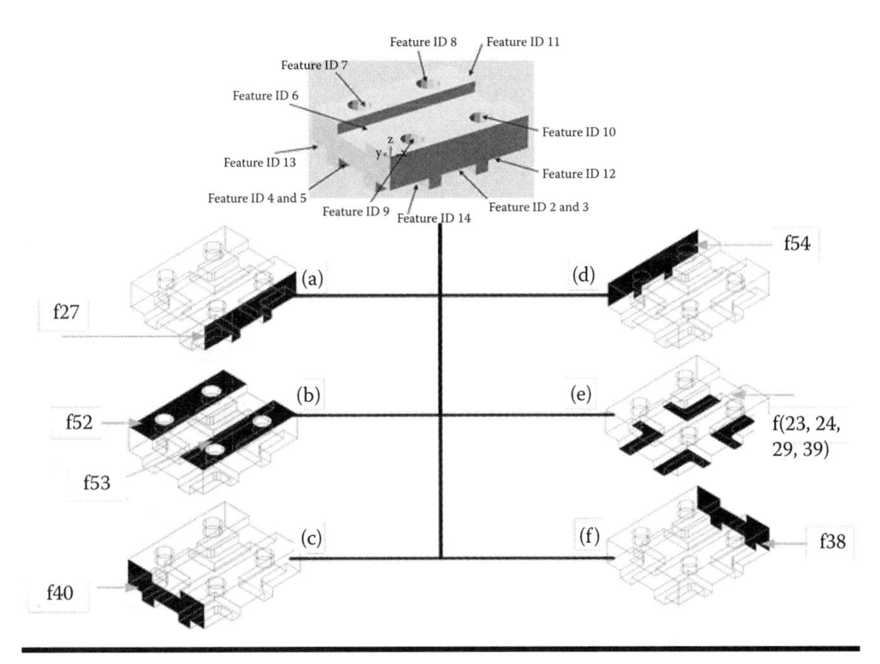

Figure 9.19 Various alternatives for the bottom (or the primary) face. Primary locating face—(a) Front (f27); (b) top (f52, f53); (c) left (f40); (d) rear (f54); (e) bottom (f23,f24, f29, f39); and (f) right (f38).

9.3.2.2 Validation of the Result of the Setup Rules

This section describes the difference between the best setup orientations and the worst setup orientation. This validation satisfied the orientation selection by avoiding increasing time in the inspection operation.

9.3.2.2.1 The Best Setup

At the front setup as shown in Figure 9.20, the matrices of the PAD and ADD of the right setup are generated automatically as listed in Tables 9.17 and 9.18. The ADD matrix gives the required dimension for the probe selection such as the length of the probe, the diameter of the probe sphere, and the nearest features to the clamp of the fixture in the CMM machine, which cannot be reached.

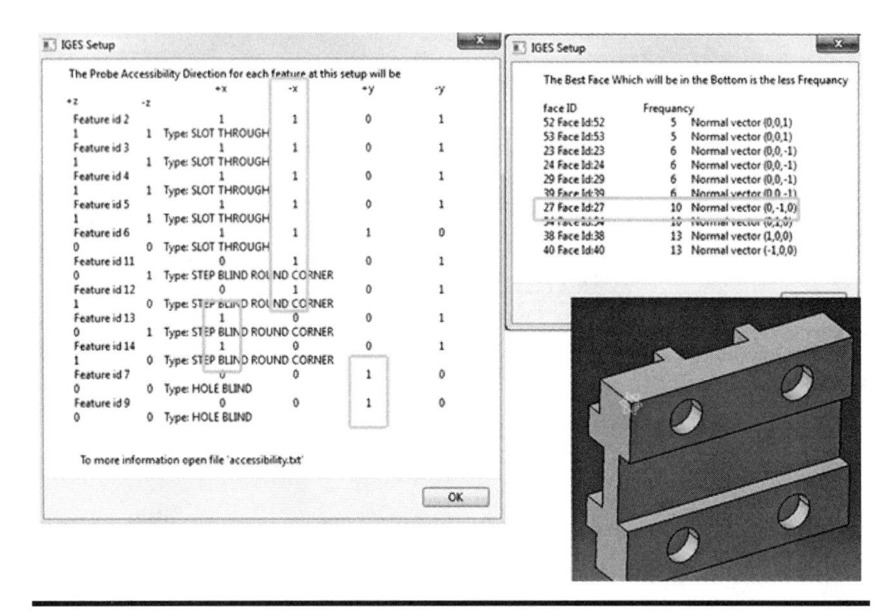

Figure 9.20 Part setups and orientation.

Table 9.17 The PAD at Right Setup

	+X	−X	+Y	−Y	+Z	−Z	
ID 2	0	1	0	1	0	1	Slot Through
ID 3	1	0	0	1	0	1	Slot Through
ID 4	0	1	0	1	1	0	Slot Through
ID 5	1	0	0	1	1	0	Slot Through
ID 6	0	0	1	0	1	1	Slot Through
ID 11	0	1	0	1	1	0	Step Blind Round Corner
ID 12	1	0	0	1	1	0	Step Blind Round Corner
ID 13	0	1	0	1	0	1	Step Blind Round Corner
ID 14	1	0	0	1	0	1	Step Blind Round Corner
ID 7	0	0	1	0	0	0	Hole Blind
ID 8	0	0	1	0	0	0	Hole Blind
ID 9	0	0	1	0	0	0	Hole Blind
ID 10	0	0	1	0	0	0	Hole Blind

Table 9.18 The ADD at Right Setup

	+X	−X	+Y	−Y	+Z	−Z	
ID 2	0	37.5	0	72.5	0	10	Slot Through
ID 3	37.5	0	0	72.5	0	10	Slot Through
ID 4	0	37.5	0	72.5	10	0	Slot Through
ID 5	37.5	0	0	72.5	10	0	Slot Through
ID 6	0	0	40	0	10	10	Slot Through
ID 11	0	10	0	40	20	0	Step Blind Round Corner
ID 12	10	0	0	40	20	0	Step Blind Round Corner
ID 13	0	10	0	40	0	20	Step Blind Round Corner
ID 14	10	0	0	40	0	20	Step Blind Round Corner
ID 7	0	0	10	0	0	0	Hole Blind
ID 8	0	0	10	0	0	0	Hole Blind
ID 9	0	0	10	0	0	0	Hole Blind
ID 10	0	0	10	0	0	0	Hole Blind

From Tables 9.17 and 9.18, the selected probe type for CMM is star probe 60 mm × 3.0 mm in diameter. The star probe can access in (−x) direction for feature IDs (2, 3, 4, 5, 6, 7, and 8), (+x) direction for feature IDs (5, 6, 7, 8, 9, 10, and 11) as shown in Figure 9.21.

9.3.2.2.2 The Worst Setup

At the top setup in Figure 9.22, the matrices of the PAD are generated automatically.

The probe selected for the CMM operation is the probe with dimensions 60 mm × 3.0 mm diameter. The probe can access in (−z) direction for feature IDs (2, 3, 4, 5, 11, 12, 13, and 14). The feature IDs (6, 7, 8, 9, and 10) cannot be reached by the probe because of the CMM fixture system as shown in Figure 9.23.

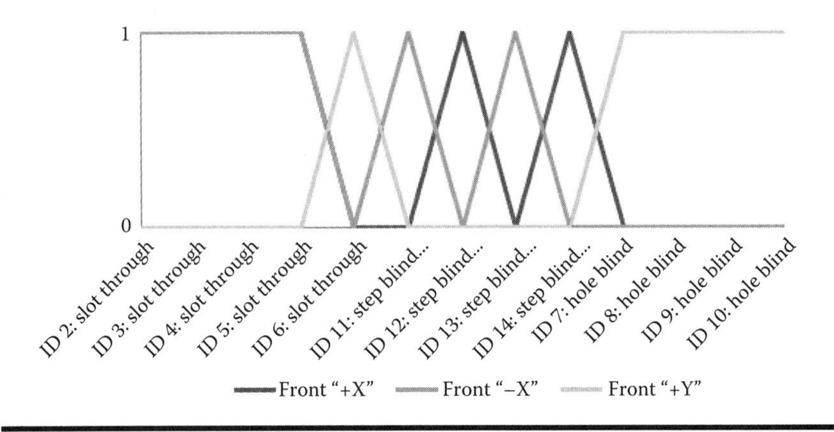

Figure 9.21 The PAD at front setup.

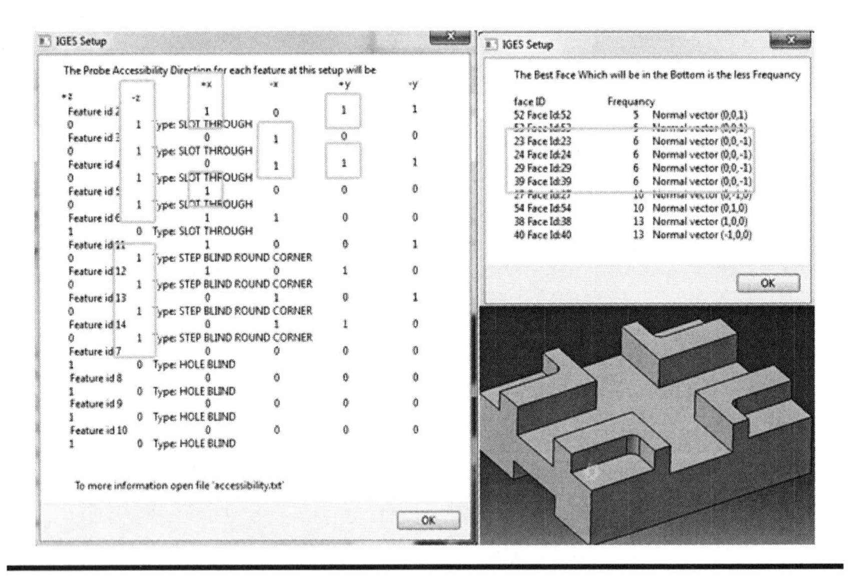

Figure 9.22 The worst (top) setup and orientation.

9.3.3 Generated Inspection Table

The generated inspection plan of case study is listed in Table 9.19 for the first and second setup of the part.

9.3.4 DMIS Code Programming

Appendix A1.3 presents the contents of the DMIS code programming generated to run CMM for the first case study.

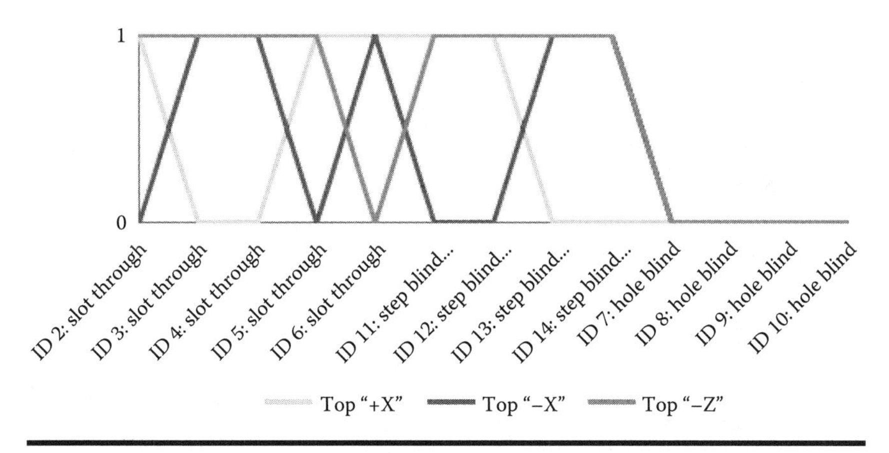

Figure 9.23 The PAD at top setup.

9.4 Illustrative Example 4 (Real Case Study)

9.4.1 Hub

9.4.1.1 GD&T Extraction

The application of the proposed methodology has been applied to the hub (mechanical pat) as shown in Figure 9.24a and b.

The proposed integrated system is used for the extraction of the GD&T from STEP file format and then generating the inspection plan shown in Table 9.20. After that, the DMIS programming code is generated using the inspection planning table. Finally, the DMIS programming code is exported to CMM.

9.4.1.2 Generated Inspection Table

9.4.1.3 DMIS Code Programming

Appendices A1.4 and A1.5 present the contents of the DMIS code programming generated to run CMM for the fourth case study at the first setup and the second setup, respectively.

Table 9.19 Inspection Planning for the First and Second Setup of the Part

Face ID	*ID of Inspection Operation*	*Tolerance Value*	*Tool Used*	*Datum Faces ID*	*Setup of the Part*	*No. of Touch Point*	*Coordinates of Touch Point with Direction*	*Geometric Boundary of Inspection Face*
AA		—	Star probe	—	First setup	6	133,82,0 133,103,0 7,103,0 7,68,0 126,89,0 126,96,0	140,75,0 140,110,0 0,110,0 0,75,0
SlMidDp		0.05	Star probe	—	First setup	6	133,42,–10 133,68,–10 7,68,–10 7,28,–10 126,49,–10 126,61,–10	0,35,–10 140,35,–10 140,75,–10 0,75,–10
		0.05	Star probe	—	First setup	6	133,82,0 133,103,0 7,103,0 7,68,0 126,89,0 126,96,0	140,75,0 140,110,0 0,110,0 0,75,0

(*Continued*)

Table 9.19 (*Continued*) Inspection Planning for the First and Second Setup of the Part

Face ID	*ID of Inspection Operation*	*Tolerance Value*	*Tool Used*	*Datum Faces ID*	*Setup of the Part*	*No. of Touch Point*	*Coordinates of Touch Point with Direction*	*Geometric Boundary of Inspection Face*
SlMidLg		0.05	Star probe	—	First setup	6	140,117,–23	140,110,–30
							140,103,7	140,110,0
							140,68,–7	140,75,0
							140,68,–3	140,75,–10
							140,42,–3	140,35,–10
							140,28,7	140,35,0
							140,–7,–7	140,0,0
							140,7,–37	140,0,–30
							140,83,–37	140,90,–30
							140,83,–33	140,90,–40
							140,79,–33	140,72,–40
							140,65,–18	140,72,–25
							140,30,–33	140,37,–25
							140,27,–33	140,37,–40
							140,117,–23	140,20,–40
							140,103,7	140,20,–30

(*Continued*)

Table 9.19 (*Continued*) Inspection Planning for the First and Second Setup of the Part

Face ID	*ID of Inspection Operation*	*Tolerance Value*	*Tool Used*	*Datum Faces ID*	*Setup of the Part*	*No. of Touch Point*	*Coordinates of Touch Point with Direction*	*Geometric Boundary of Inspection Face*
		0.05	Star probe	—	First setup	8	0,117,–7	0,110,0
							0,103,–37	0,110,–30
							0,97,–37	0,90,–30
							0,83,–47	0,90,–40
							0,65,–33	0,72,–40
							0,65,–32	0,72,–25
							0,44,–32	0,37,–25
							0,30,–47	0,37,–40
							0,13,–33	0,20,–40
							0,13,–37	0,20,–30
							0,–7,–23	0,0,–30
							0,7,7	0,0,0
							0,82,–7	0,75,0
							0,68,–17	0,75,–10
							0,28,–3	0,35,–10
							0,117,–7	0,35,0

(*Continued*)

Table 9.19 (*Continued*) Inspection Planning for the First and Second Setup of the Part

Face ID	*ID of Inspection Operation*	*Tolerance Value*	*Tool Used*	*Datum Faces ID*	*Setup of the Part*	*No. of Touch Point*	*Coordinates of Touch Point with Direction*	*Geometric Boundary of Inspection Face*
FrHolCy	⌭	0.001	Star probe	—	First setup	8	45,17.5,−1,1,0,0 35,27.5,−1,0,−1,0 25,17.5,−1,−1,0,0 35,7.5,−1,0,1,0 45,17.5,−2,1,0,0 35,27.5,−2,0,−1,0 25,17.5,−2,−1,0,0 35,7.5,−2,0,1,0	Center point 35.,17.5,−5 — Radius 10 — Height 10
FrHolCy	○	0.001	Star probe	—	First setup	8	45,17.5,−1,1,0,0 35,27.5,−1,0,−1,0 25,17.5,−1,−1,0,0 35,7.5,−1,0,1,0 45,17.5,−2,1,0,0 35,27.5,−2,0,−1,0 25,17.5,−2,−1,0,0 35,7.5,−2,0,1,0	Center point 35.,17.5,−5 — Radius 10 — Height 10

(*Continued*)

Table 9.19 (*Continued*) Inspection Planning for the First and Second Setup of the Part

Face ID	*ID of Inspection Operation*	*Tolerance Value*	*Tool Used*	*Datum Faces ID*	*Setup of the Part*	*No. of Touch Point*	*Coordinates of Touch Point with Direction*	*Geometric Boundary of Inspection Face*
ScHolCy	⌭	0.001	Star probe	—	First setup	8	115,17.5,–1,1,0,0 105,27.5,–1,0,–1,0 95,17.5,–1,–1,0,0 105,7.5,–1,0,1,0 115,17.5,–2,1,0,0 105,27.5,–2,0,–1,0 95,17.5,–2,–1,0,0 105,7.5,–2,0,1,0	Center point 35.,17.5,–5 — Radius 10 — Height 10
ThHolCy	⌭	0.001	Star probe	—	First setup	8	45,92.5,–1,1,0,0 35,102.5,–1,0,–1,0 25,92.5,–1,–1,0,0 35,82.5,–1,0,1,0 45,92.5,–2,1,0,0 35,102.5,–2,0,–1,0 25,92.5,–2,–1,0,0 35,82.5,–2,0,1,0	Center point 35.,92.5,–5 — Radius 10 — Height 10

(*Continued*)

Table 9.19 (*Continued*) Inspection Planning for the First and Second Setup of the Part

Face ID	*ID of Inspection Operation*	*Tolerance Value*	*Tool Used*	*Datum Faces ID*	*Setup of the Part*	*No. of Touch Point*	*Coordinates of Touch Point with Direction*	*Geometric Boundary of Inspection Face*
FuHolCy	⌭	0.001	Star probe	—	First setup	8	115,92.5,−1,1,0,0 105,102.5,−1,0,−1,0 95,92.5,−1,−1,0,0 105,82.5,−1,0,1,0 115,92.5,−2,1,0,0 105,102.5,−2,0,−1,0 95,92.5,−2,−1,0,0 105,82.5,−2,0,1,0	Center point 105.,92.5,−5 — Radius 10 — Height 10
SlMidFr	⊥	0.05	Star probe	AA	First setup	6	133,75,−3 7,75,−3 7,75,−7 133,75,7 126,75,4 0,75,−10	140,75,0 140,75,−10 0,75,−10 0,75,0
SlMidBo	▱	0.002	Star probe	AA	First setup	6	133,42,−10 133,68,−10 7,68,−10 7,28,−10 126,49,−10 126,61,−10	0,35,−10 140,35,−10 140,75,−10 0,75,−10

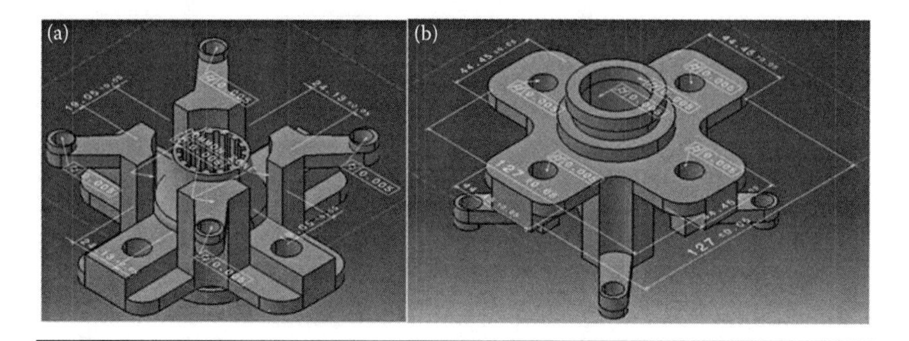

Figure 9.24 (a) Hub in the first setup and (b) hub in the second setup.

9.4.2 Gear Pump Housing

9.4.2.1 GD&T Extraction

The application of the integrated system has also been applied to the gear pump housing (mechanical pat) as shown in Figure 9.25a and b.

The proposed system is used for extraction of the GD&T from STEP file format and then generating the inspection plan shown in Table 9.21. After that, the DMIS programming code is generated using the inspection planning table. Finally, the DMIS programming code is exported to the CMM.

9.4.2.2 Generated Inspection Table

9.4.2.3 DMIS Code Programming

Appendices A1.6 and A1.7 present the contents of the DMIS code programming generated to run CMM for the Fourth case study at the first setup and the second setup, respectively.

9.5 Summary

In this chapter, the case studies consisting of prismatic parts and real manufacturing parts (hub and gear pump housing) have successfully been carried out to show the feasibility and applications of the integrated system of CAD and inspection planning.

Table 9.20 Inspection Plan for Actual Part (Hub)

Face ID	*ID of Inspection Operation*	*Tolerance Value*	*Tool Used*	*Datum Faces ID*	*Setup of the Part*	*No. of Touch Point*	*Coordinates of Touch Point with Direction*	*Geometric Boundary of Inspection Face*
FrHole	⌭	0.005	Star probe	—	First setup	8	–57.5,–107.95,–1.5875,1,0,0 –63.5,–101.95,–1.5875,0,–1,0 –69.5,-107.95,–1.5875,–1,0,0 –63.5,–113.95,–1.5875,0,1,0 –57.5,–107.95,–3.175,1,0,0 –63.5,–101.95,–3.175,0,–1,0 –69.5,–107.95,–3.175,–1,0,0 –63.5,–113.95,–3.175,0,1,0	Center point –63.5,–107.95,–30.1625 Radius 6 Height 15.875
ScHole	⌭	0.005	Star probe	—	First setup	8	–101.95,–63.5,–1.5875,1,0,0 –107.95,–57.5,–1.5875,0,–1,0 –113.95,–63.5,–1.5875,–1,0,0 –107.95,–69.5,–1.5875,0,1,0 –101.95,–63.5,–3.175,1,0,0 –107.95,–57.5,–3.175,0,–1,0 –113.95,–63.5,–3.175,–1,0,0 –107.95,–69.5,–3.175,0,1,0	Center point –107.95,–63.5,–30.1625 Radius 6 Height 15.875

(Continued)

Table 9.20 (*Continued*) Inspection Plan for Actual Part (Hub)

Face ID	*ID of Inspection Operation*	*Tolerance Value*	*Tool Used*	*Datum Faces ID*	*Setup of the Part*	*No. of Touch Point*	*Coordinates of Touch Point with Direction*	*Geometric Boundary of Inspection Face*
TrHole	⌭	0.005	Star probe	—	First setup	8	−57.5,−19.05,−1.5875,1,0,0 −63.5,−13.05,−1.5875,0,−1,0 −69.5,−19.05,−1.5875,−1,0,0 −63.5,−25.05,−1.5875,0,1,0 −57.5,−19.05,−3.175,1,0,0 −63.5,−13.05,−3.175,0,−1,0 −69.5,−19.05,−3.175,−1,0,0 −63.5,−25.05,−3.175,0,1,0	Center point −63.5,−19.05,−30.1625 Radius 6 Height 15.875
FuHole	⌭	0.005	Star probe	—	First setup	8	−13.05,−63.5,−1.5875,1,0,0 −19.05,−57.5,−1.5875,0,−1,0 −25.05,−63.5,−1.5875,−1,0,0 −19.05,−69.5,−1.5875,0,1,0 −13.05,−63.5,−3.175,1,0,0 −19.05,−57.5,−3.175,0,−1,0 −25.05,−63.5,−3.175,−1,0,0 −19.05,−69.5,−3.175,0,1,0	Center point −19.05,−63.5,−30.1625 Radius 6 Height 15.875

(*Continued*)

Table 9.20 (*Continued*) Inspection Plan for Actual Part (Hub)

Face ID	*ID of Inspection Operation*	*Tolerance Value*	*Tool Used*	*Datum Faces ID*	*Setup of the Part*	*No. of Touch Point*	*Coordinates of Touch Point with Direction*	*Geometric Boundary of Inspection Face*
UMHole	⌭	0.004	Star probe	—	First setup	8	−48.5,−63.5,−0.635,1,0,0 −63.5,−48.5,−0.635,0,−1,0 −78.5,−63.5,−0.635,−1,0,0 −63.5,−78.5,−0.635,0,1,0 −48.5,−63.5,−1.27,1,0,0 −63.5,−48.5,−1.27,0,−1,0 −78.5,−63.5,−1.27,−1,0,0 −63.5,−78.5,−1.27,0,1,0	Center point −63.5,−63.5,−3.175 Radius 15 Height 6.350
DMHole	⌭	0.004	Star probe	—	First setup	8	−48.5,−63.5,−1.9,1,0,0 −63.5,−48.5,−1.9,0,−1,0 −78.5,−63.5,−1.9,−1,0,0 −63.5,−78.5,−1.9,0,1,0 −48.5,−63.5,−3.8,1,0,0 −63.5,−48.5,−3.8,0,−1,0 −78.5,−63.5,−3.8,−1,0,0 −63.5,−78.5,−3.8,0,1,0	Center point −63.5,−63.5,−22.225 Radius 15 Height 19.05

(*Continued*)

Table 9.20 (*Continued*) Inspection Plan for Actual Part (Hub)

Face ID	*ID of Inspection Operation*	*Tolerance Value*	*Tool Used*	*Datum Faces ID*	*Setup of the Part*	*No. of Touch Point*	*Coordinates of Touch Point with Direction*	*Geometric Boundary of Inspection Face*
HUBWdth		0.05	Star probe	—	First setup	6	−127,−56,−29 −127,−70,−29 −127,−70,−24 −127,−61,−24 −127,−65,−24 −127,−70,−24	−127,−75,−34 −127,−75,−19 −127,−51,−19 −127,−51,−34
		0.05	Star probe	—	First setup	8	3,−56,−29 3,−56,−24 3,−70,−24 3,−61,−24 3,−61,−29 3,−70,−24	3,−51,−34 3,−51,−19 3,−75,−19 3,−75,−34
HUBLnth		0.05	Star probe	—	First setup	10	−58,−127,−29 −58,−127,−24 −68,−127,−24 −68,−127,−29 −63,−127,−24 −63,−127,−29	−53,−127,−34 −53,−127,−19 −73,−127,−19 −73,−127,−34

(*Continued*)

Table 9.20 (*Continued*) Inspection Plan for Actual Part (Hub)

Face ID	*ID of Inspection Operation*	*Tolerance Value*	*Tool Used*	*Datum Faces ID*	*Setup of the Part*	*No. of Touch Point*	*Coordinates of Touch Point with Direction*	*Geometric Boundary of Inspection Face*
		0.05	Star probe	—	First setup	6	−58,2,−29 −68,2,−29 −68,2,−24 −58,2,−24 −63,2,−24 −63,2,−24	−73,2,−34 −73,2,−19 −53,2,−19 −53,2,−34
FuLenth		0.05	Star probe	—	First setup	6	−15,−85,−20 −26,−85,−20 −26,−85,−24 −15,−85,−24 −20,−85,−15 −31,−85,−25	−31,−85,−25 −31,−85,−19 −10,−85,−19 −10,−85,−25
		0.05	Star probe	—	First setup	6	−15,−41,−20 −15,−41,−24 −26,−41,−24 126,−41,−20 −10,−41,−25 −20,−41,−29	−10,−41,−25 −10,−41,−19 −31,−41,−19 −31,−41,−25

(*Continued*)

Table 9.20 (*Continued*) Inspection Plan for Actual Part (Hub)

Face ID	*ID of Inspection Operation*	*Tolerance Value*	*Tool Used*	*Datum Faces ID*	*Setup of the Part*	*No. of Touch Point*	*Coordinates of Touch Point with Direction*	*Geometric Boundary of Inspection Face*
TrLenth		0.05	Star probe	—	Second setup	6	−85,−17,−20 −85,−17,−24 −85,−26,−24 −85,−12,−25 −85,−22,−29 −85,−26,−24	−85,−12,−25 −85,−12,−19 −85,−31,−19 −85,−31,−25
		0.05	Star probe	—	Second setup	6	−41,−17,−20 −41,−26,−20 −41,−26,−24 −41,−22,−15 −41,−31,−25 −41,−26,−24	−41,−31,−25 −41,−31,−19 −41,−12,−19 −41,−12,−25
ScLenth		0.05	Star probe	—	Second setup	6	−100,−85,−20 −111,−85,−20 −111,−85,−24 −100,−85,−24 −105,−85,−15 −116,−85,−25	−116,−85,−25 −116,−85,−19 −95,−85,−19 −95,−85,−25

(*Continued*)

Table 9.20 (*Continued*) Inspection Plan for Actual Part (Hub)

Face ID	*ID of Inspection Operation*	*Tolerance Value*	*Tool Used*	*Datum Faces ID*	*Setup of the Part*	*No. of Touch Point*	*Coordinates of Touch Point with Direction*	*Geometric Boundary of Inspection Face*
		0.05	Star probe	—	Second setup	6	−100,−41,−20 −100,−41,−24 −111,−41,−24 −111,−41,−20 −95,−41,−25 −105,−41,−29	−95,−41,−25 −95,−41,−19 −116,−41,−19 −116,−41,−25
FrLenth		0.05	Star probe	—	Second setup	6	−41,−100,−20 −41,−109,−20 −41,−109,−24 −41,−105,−15 −41,−114,−25 −41,−109,−24	−41,−114,−25 −41,−114,−19 −41,−95,−19 −41,−95,−25
		0.05	Star probe	—	Second setup	6	−85,−100,−20 −85,−100,−24 −85,−109,−24 −85,−95,−25 −85,−105,−29 −85,−109,−24	−85,−95,−25 −85,−95,−19 −85,−114,−19 −85,−114,−25

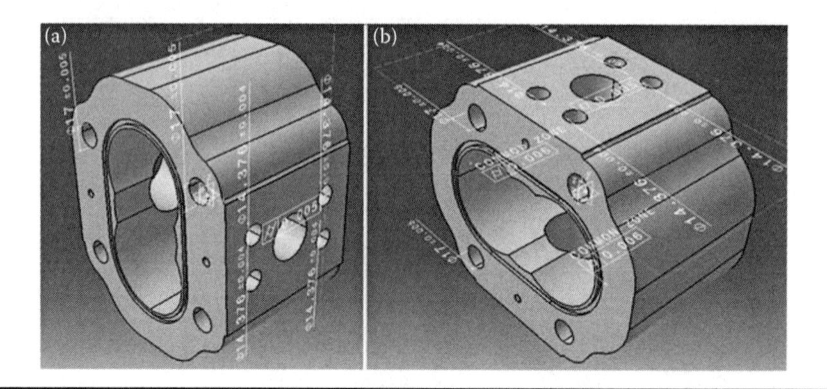

Figure 9.25 (a) Gear pump housing in the first setup and (b) gear pump housing in the second setup.

QUESTIONS

1. What machining information can be associated with the designed part?
2. Write down the information that can be extracted from the manufacturing features.
3. What do you mean by probe approach direction (PAD)?
4. What information is included in the generated inspection plan table?
5. Write down the benefits of choosing the right set up for the given part.
6. Describe the validation process of the setup rules.
7. How is the integration of CAD and inspection planning beneficial to the manufacturing industry?
8. Write down the different steps for the implementation of the integrated system to the following components:
   a. Hub
   b. Gear pump housing

Table 9.21 Inspection Plan for Actual Part (Gear Pump Housing)

Face ID	*ID of Inspection Operation*	*Tolerance Value*	*Tool Used*	*Datum Faces ID*	*Setup of the Part*	*No. of Touch Point*	*Coordinates of Touch Point with Direction*	*Geometric Boundary of Inspection Face*
FrHoleD	⌭	0.005	Star probe	—	First setup	8	7,48.7,0,1,0,0 0,48.7,7,0,0,1 −7,48.7,0,−1,0,0 0,48.7,−7,0,0,−1 7,51.9,0,1,0,0 0,51.9,7,0,0,1 −7,51.9,0,−1,0,0 0,51.9,−7,0,0,−1	Center point 0,77.4,0 Radius 7 Height 32
ScHoleD	⌭	0.005	Star probe	—	First setup	8	7,48.7,15.9,1,0,0 0,48.7,22.9,0,0,1 −7,48.7,15.9,−1,0,0 0,48.7,8.9,0,0,−1 7,51.9,15.9,1,0,0 0,51.9,22.9,0,0,1 −7,51.9,15.9,−1,0,0 0,51.9,8.9,0,0,−1	Center point 0,77.4,15.9 Radius 7 Height 32

(*Continued*)

Table 9.21 (*Continued*) Inspection Plan for Actual Part (Gear Pump Housing)

Face ID	*ID of Inspection Operation*	*Tolerance Value*	*Tool Used*	*Datum Faces ID*	*Setup of the Part*	*No. of Touch Point*	*Coordinates of Touch Point with Direction*	*Geometric Boundary of Inspection Face*
ThHoleD	⌭	0.005	Star probe	—	First setup	8	40.35,48.7,0,1,0,0 33.35,48.7,7,0,0,1 26.35,48.7,0,–1,0,0 33.35,48.7,–7,0,0,–1 40.35,51.9,0,1,0,0 33.35,51.9,7,0,0,1 26.35,51.9,0,–1,0,0 33.35,51.9,–7,0,0,–1	Center point 33.35,77.4,0 Radius 7 Height 32
FuHoleD	⌭	0.005	Star probe	—	First setup	8	–40.35,48.7,15.9,1,0,0 33.35,48.7,22.9,0,0,1 26.35,48.7,15.9,–1,0,0 33.35,48.7,8.9,0,0,–1 40.35,51.9,15.9,1,0,0 33.35,51.9,22.9,0,0,1 26.35,51.9,15.9,–1,0,0 33.35,51.9,8.9,0,0,–1	Center point 33.35,77.4, 15.9 Radius 7 Height 32

(*Continued*)

Table 9.21 (*Continued*) Inspection Plan for Actual Part (Gear Pump Housing)

Face ID	*ID of Inspection Operation*	*Tolerance Value*	*Tool Used*	*Datum Faces ID*	*Setup of the Part*	*No. of Touch Point*	*Coordinates of Touch Point with Direction*	*Geometric Boundary of Inspection Face*
LfFHolD	⌭	0.004	Star probe	—	Second setup	8	45.16,8,45,0,1,0 45.16,0,53,0,0,1 45.16,–8,45,0,–1,0 45.16,0,37,0,0,1 33.87,8,45,0,1,0 33.87,0,53,0,0,1 33.87,–8,45,0,–1,0 33.87,0,37,0,0,1	Center point 28.225,0,45 Radius 8 Height 112.9
RgFHolD	⌭	0.004	Star probe	—	Second setup	8	–45.16,63,45,0,1,0 –45.16,55,53,0,0,1 –45.16,47,45,0,–1,0 –45.16,55,37,0,0,–1 –33.87,63,45,0,1,0 –33.87,55,53,0,0,1 –33.87,47,45,0,—1,0 –33.87,55,37,0,0,–1	Center point 28.225,55.,45 Radius 15 Height 19.05

Appendix: DMIS Code Programming

A1.1 DMIS Code Programming of First Case Study

```
$$ CARL ZEISS – CALYPSO Preprocessor
$$ Ver.1.01.029.00 Date: Mon Dec 17 2012 Time: 11:35:44

DMISMN/'Case study No. 1'
FILNAM/'CaseFour.DMI'
UNITS/MM, ANGDEC, TEMPC

SNSET/APPRCH, 2.0000
SNSET/RETRCT, 5
SNSET/SEARCH, 5.0000

$$ Changing label 'Star ~ 1' to: 'STAR1'.
S(STAR1)=SNSDEF/PROBE, INDEX, CART, –0.0150, 0.1975,
–20.9113, 0.0000, 0.0000, $
–1.0000, 3.0050, SPHERE

RECALL/DA(REFERENCE)

MODE/PROG,MAN
```

```
F(AA_0)=FEAT/PLANE, CART, 55,110,0,0,0,1
MEAS / PLANE, F(AA_0), 3
PTMEAS/CART, 51,4,0,0,0,1
PTMEAS/CART, 51,106,0,0,0,1
PTMEAS/CART, 19,106,0,0,0,1

GOTO/19,106,20
GOTO/19,106,20
GOTO/10,106,20
GOTO/10,106,–11
ENDMES

F(StOneL_1)=FEAT/PLANE, CART, 15,0,0,–1,0,0
MEAS / PLANE, F(StOneL_1), 3
PTMEAS/CART, 15,106,–11,–1,0,0
PTMEAS/CART, 15,106,–4,–1,0,0
PTMEAS/CART, 15,4,–4,–1,0,0

GOTO/10,4,–4
GOTO/10,4,20
GOTO/10,106,20
GOTO/–5,106,20
GOTO/–5,106,–36
ENDMES

F(StOneL_2)=FEAT/PLANE, CART, 0,110,–20,–1,0,0
MEAS / PLANE, F(StOneL_2), 3
PTMEAS/CART, 0,106,–36,–1,0,0
PTMEAS/CART, 0,79,–36,–1,0,0
PTMEAS/CART, 0,71,–21,–1,0,0

GOTO/–5,71,–21
GOTO/–5,71,20
GOTO/–5,4,20
GOTO/51,4,20
```

```
GOTO/51,4,20
ENDMES

F(StOneDp_3)=FEAT/PLANE, CART, 55,110,0,0,0,1
MEAS / PLANE, F(StOneDp_3), 3
PTMEAS/CART, 51,4,0,0,0,1
PTMEAS/CART, 51,106,0,0,0,1
PTMEAS/CART, 19,106,0,0,0,1

GOTO/19,106,20
GOTO/19,4,20
GOTO/6,4,20
GOTO/6,4,0
ENDMES

F(StOneDp_4)=FEAT/PLANE, CART, 10,110,–20,0,0,1
MEAS / PLANE, F(StOneDp_4), 3
PTMEAS/CART, 6,4,–20,0,0,1
PTMEAS/CART, 6,106,–20,0,0,1
PTMEAS/CART, 4,106,–20,0,0,1

GOTO/4,106,20
GOTO/4,106,20
GOTO/60,106,20
GOTO/60,106,–11
ENDMES

F(StMiddL_5)=FEAT/PLANE, CART, 55,110,0,1,0,0
MEAS / PLANE, F(StMiddL_5), 3
PTMEAS/CART, 55,106,–11,1,0,0
PTMEAS/CART, 55,4,–11,1,0,0
PTMEAS/CART, 55,4,–4,1,0,0

GOTO/60,4,–4
GOTO/60,4,20
GOTO/60,106,20
```

```
GOTO/80,106,20
GOTO/80,106,–11
ENDMES

F(StMiddL_6)=FEAT/PLANE, CART, 85,0,0,–1,0,0
MEAS / PLANE, F(StMiddL_6), 3
PTMEAS/CART, 85,106,–11,–1,0,0
PTMEAS/CART, 85,106,–4,–1,0,0
PTMEAS/CART, 85,4,–4,–1,0,0

GOTO/80,4,–4
GOTO/80,4,20
GOTO/80,4,20
GOTO/121,4,20
GOTO/121,4,20
ENDMES

F(StMidDp_7)=FEAT/PLANE, CART, 85,0,0,0,0,1
MEAS / PLANE, F(StMidDp_7), 3
PTMEAS/CART, 121,4,0,0,0,1
PTMEAS/CART, 121,106,0,0,0,1
PTMEAS/CART, 89,106,0,0,0,1

GOTO/89,106,20
GOTO/89,4,20
GOTO/76,4,20
GOTO/76,4,0
ENDMES

F(StMidDp_8)=FEAT/PLANE, CART, 60,0,–20,0,0,1
MEAS / PLANE, F(StMidDp_8), 3
PTMEAS/CART, 76,4,–20,0,0,1
PTMEAS/CART, 76,106,–20,0,0,1
PTMEAS/CART, 64,106,–20,0,0,1
GOTO/64,106,20
GOTO/64,4,20
```

```
GOTO/121,4,20
GOTO/121,4,20
ENDMES

F(StTwoDp_9)=FEAT/PLANE, CART, 85,0,0,0,0,1
MEAS / PLANE, F(StTwoDp_9), 3
PTMEAS/CART, 121,4,0,0,0,1
PTMEAS/CART, 121,106,0,0,0,1
PTMEAS/CART, 89,106,0,0,0,1

GOTO/89,106,20
GOTO/89,4,20
GOTO/136,4,20
GOTO/136,4,0
ENDMES

F(StTwoDp_10)=FEAT/PLANE, CART, 130,0,–20,0,0,1
MEAS / PLANE, F(StTwoDp_10), 3
PTMEAS/CART, 136,4,–20,0,0,1
PTMEAS/CART, 136,106,–20,0,0,1
PTMEAS/CART, 134,106,–20,0,0,1

GOTO/134,106,20
GOTO/134,106,20
GOTO/130,106,20
GOTO/130,106,–11
ENDMES

F(StTwoL_11)=FEAT/PLANE, CART, 125,0,–15,1,0,0
MEAS / PLANE, F(StTwoL_11), 3
PTMEAS/CART, 125,106,–11,1,0,0
PTMEAS/CART, 125,106,–4,1,0,0
PTMEAS/CART, 125,4,–4,1,0,0

GOTO/130,4,–4
GOTO/130,4,20
GOTO/130,106,20
```

```
GOTO/145,106,20
GOTO/145,106,–36
ENDMES

F(StTwoL_12)=FEAT/PLANE, CART, 140,110,–40,1,0,0
MEAS / PLANE, F(StTwoL_12), 3
PTMEAS/CART, 140,106,–36,1,0,0
PTMEAS/CART, 140,106,–24,1,0,0
PTMEAS/CART, 140,4,–24,1,0,0

GOTO/145,4,–24
GOTO/145,4,20
GOTO/145,106,20
GOTO/80,106,20
GOTO/80,106,–11
ENDMES

F(StMidRg_13)=FEAT/PLANE, CART, 85,0,0,–1,0,0
MEAS / PLANE, F(StMidRg_13), 3
PTMEAS/CART, 85,106,–11,–1,0,0
PTMEAS/CART, 85,106,–4,–1,0,0
PTMEAS/CART, 85,4,–4,–1,0,0

GOTO/80,4,–4
GOTO/80,4,20
GOTO/80,4,20
GOTO/76,4,20
GOTO/76,4,0
ENDMES

F(StMidBt_14)=FEAT/PLANE, CART, 60,0,–20,0,0,1
MEAS / PLANE, F(StMidBt_14), 3
PTMEAS/CART, 76,4,–20,0,0,1
PTMEAS/CART, 76,106,–20,0,0,1
PTMEAS/CART, 64,106,–20,0,0,1
```

```
GOTO/64,106,20
GOTO/64,106,20
GOTO/60,106,20
GOTO/60,106,–11
ENDMES

F(StMidLf_15)=FEAT/PLANE, CART, 55,110,0,1,0,0
MEAS / PLANE, F(StMidLf_15), 3
PTMEAS/CART, 55,106,–11,1,0,0
PTMEAS/CART, 55,4,–11,1,0,0
PTMEAS/CART, 55,4,–4,1,0,0

GOTO/60,4,–4
GOTO/60,4,20
GOTO/60,17.5,20
GOTO/27.5,17.5,20
GOTO/27.5,17.5,20
ENDMES

F(FrHole_16)=FEAT/CYLNDR, INNER, CART,
27.5,17.5,–20.0,0,0,1,20,–40
MEAS / CYLNDR, F(FrHole_16), 12
PTMEAS/CART, 37.5,17.5,–4,1,0,0
PTMEAS/CART, 27.5,27.5,–4,0,–1,0
PTMEAS/CART, 17.5,17.5,–4,–1,0,0
PTMEAS/CART, 27.5,7.5,–4,0,1,0
PTMEAS/CART, 37.5,17.5,–8,1,0,0
PTMEAS/CART, 27.5,27.5,–8,0,–1,0
PTMEAS/CART, 17.5,17.5,–8,–1,0,0
PTMEAS/CART, 27.5,7.5,–8,0,1,0
PTMEAS/CART, 37.5,17.5,–12,1,0,0
PTMEAS/CART, 27.5,27.5,–12,0,–1,0
PTMEAS/CART, 17.5,17.5,–12,–1,0,0
PTMEAS/CART, 27.5,7.5,–12,0,1,0
```

```
GOTO/27.5,17.5,20
GOTO/27.5,17.5,20
GOTO/112.5,17.5,20
GOTO/112.5,17.5,20
ENDMES

F(ScHole_17)=FEAT/CYLNDR, INNER, CART,
112.5,17.5,–20.0,0,0,1,20,–40
MEAS / CYLNDR, F(ScHole_17), 12
PTMEAS/CART, 122.5,17.5,–4,1,0,0
PTMEAS/CART, 112.5,27.5,–4,0,–1,0
PTMEAS/CART, 102.5,17.5,–4,–1,0,0
PTMEAS/CART, 112.5,7.5,–4,0,1,0
PTMEAS/CART, 122.5,17.5,–8,1,0,0
PTMEAS/CART, 112.5,27.5,–8,0,–1,0
PTMEAS/CART, 102.5,17.5,–8,–1,0,0
PTMEAS/CART, 112.5,7.5,–8,0,1,0
PTMEAS/CART, 122.5,17.5,–12,1,0,0
PTMEAS/CART, 112.5,27.5,–12,0,–1,0
PTMEAS/CART, 102.5,17.5,–12,–1,0,0
PTMEAS/CART, 112.5,7.5,–12,0,1,0

GOTO/112.5,17.5,20
GOTO/112.5,92.5,20
GOTO/27.5,92.5,20
GOTO/27.5,92.5,20
ENDMES

F(ThHole_18)=FEAT/CYLNDR, INNER, CART,
27.5,92.5,–20.0,0,0,1,20,–40
MEAS / CYLNDR, F(ThHole_18), 12
PTMEAS/CART, 37.5,92.5,–4,1,0,0
PTMEAS/CART, 27.5,102.5,–4,0,–1,0
PTMEAS/CART, 17.5,92.5,–4,–1,0,0
PTMEAS/CART, 27.5,82.5,–4,0,1,0
```

```
PTMEAS/CART, 37.5,92.5,–8,1,0,0
PTMEAS/CART, 27.5,102.5,–8,0,–1,0
PTMEAS/CART, 17.5,92.5,–8,–1,0,0
PTMEAS/CART, 27.5,82.5,–8,0,1,0
PTMEAS/CART, 37.5,92.5,–12,1,0,0
PTMEAS/CART, 27.5,102.5,–12,0,–1,0
PTMEAS/CART, 17.5,92.5,–12,–1,0,0
PTMEAS/CART, 27.5,82.5,–12,0,1,0

GOTO/27.5,92.5,20
GOTO/27.5,92.5,20
GOTO/112.5,92.5,20
GOTO/112.5,92.5,20
ENDMES

F(FuHole_19)=FEAT/CYLNDR, INNER, CART,
112.5,92.5,–20.0,0,0,1,20,–40
MEAS / CYLNDR, F(FuHole_19), 12
PTMEAS/CART, 122.5,92.5,–4,1,0,0
PTMEAS/CART, 112.5,102.5,–4,0,–1,0
PTMEAS/CART, 102.5,92.5,–4,–1,0,0
PTMEAS/CART, 112.5,82.5,–4,0,1,0
PTMEAS/CART, 122.5,92.5,–8,1,0,0
PTMEAS/CART, 112.5,102.5,–8,0,–1,0
PTMEAS/CART, 102.5,92.5,–8,–1,0,0
PTMEAS/CART, 112.5,82.5,–8,0,1,0
PTMEAS/CART, 122.5,92.5,–12,1,0,0
PTMEAS/CART, 112.5,102.5,–12,0,–1,0
PTMEAS/CART, 102.5,92.5,–12,–1,0,0
PTMEAS/CART, 112.5,82.5,–12,0,1,0

ENDMES

T(genera_1)=TOL/DISTB , NOMINL,15, 0.05, 0.05, PT2PT
OUTPUT/FA(StOneL_1),FA(StOneL_2), TA(genera_1)
```

```
T(genera_3)=TOL/DISTB, NOMINL,20, 0.05, 0.05, PT2PT
OUTPUT/FA(StOneDp_3),FA(StOneDp_4), TA(genera_3)

T(genera_5)=TOL/DISTB, NOMINL,30, 0.05, 0.05, PT2PT
OUTPUT/FA(StMiddL_5),FA(StMiddL_6), TA(genera_5)

T(genera_7)=TOL/DISTB, NOMINL, 20, 0.05, 0.05, PT2PT
OUTPUT/FA(StMidDp_7),FA(StMidDp_8), TA(genera_7)

T(genera_9)=TOL/DISTB, NOMINL, 20, 0.05, 0.05, PT2PT
OUTPUT/FA(StTwoDp_9),FA(StTwoDp_10), TA(genera_9)

T(genera_11)=TOL/DISTB, NOMINL, 15, 0.05, 0.05, PT2PT
OUTPUT/FA(StTwoL_11),FA(StTwoL_12), TA(genera_11)

T(flatne_13)=TOL/FLAT, 0.05
OUTPUT/FA(StMidRg_13), TA(flatne_13)

T(parall_14)=TOL/PARLEL, 0.001
OUTPUT/FA(StMidBt_14), TA(parall_14)

T(perpen_15)=TOL/PERP, 0.001,FA(AA_0)
OUTPUT/FA(StMidLf_15), TA(perpen_15)

T(cylind_16)=TOL/CYLCTY, 0.001
OUTPUT/FA(FrHole_16), TA(cylind_16)

T(circul_17)=TOL/CIRLTY, 0.001
OUTPUT/FA(ScHole_17), TA(circul_17)

T(circul_18)=TOL/CIRLTY, 0.001
OUTPUT/FA(ThHole_18), TA(circul_18)

T(cylind_19)=TOL/CYLCTY, 0.001
OUTPUT/FA(FuHole_19), TA(cylind_19)

ENDFIL
```

A1.2 DMIS Code Programming of Second Case Study

```
$$ CARL ZEISS – CALYPSO Preprocessor
$$ Ver.1.01.029.00 Date: Mon Dec 17 2012 Time: 11:35:44

DMISMN/'Case Study No. 2'
FILNAM/'CaseFour.DMI'
UNITS/MM, ANGDEC, TEMPC
SNSET/APPRCH, 2.0000
SNSET/RETRCT, 5
SNSET/SEARCH, 5.0000
$$ Changing label 'Star ~ 1' to: 'STAR1'.
S(STAR1)=SNSDEF/PROBE, INDEX, CART, –0.0150, 0.1975,
–20.9113, 0.0000, 0.0000, $
–1.0000, 3.0050, SPHERE

RECALL/DA(REFERENCE)
MODE/PROG,MAN

F(AA_0)=FEAT/PLANE, CART, 55,110,0,0,0,1
MEAS / PLANE, F(AA_0), 3
PTMEAS/CART, 51,4,0,0,0,1
PTMEAS/CART, 51,106,0,0,0,1
PTMEAS/CART, 19,106,0,0,0,1

GOTO/19,106,20
GOTO/19,106,20
GOTO/10,106,20
GOTO/10,106,–11
ENDMES

F(StOneL_1)=FEAT/PLANE, CART, 15,0,0,–1,0,0
MEAS / PLANE, F(StOneL_1), 3
PTMEAS/CART, 15,106,–11,–1,0,0
```

```
PTMEAS/CART, 15,106,–4,–1,0,0
PTMEAS/CART, 15,4,–4,–1,0,0

GOTO/10,4,–4
GOTO/10,4,20
GOTO/10,106,20
GOTO/–5,106,20
GOTO/–5,106,–36
ENDMES

F(StOneL_2)=FEAT/PLANE, CART, 0,110,–20,–1,0,0
MEAS / PLANE, F(StOneL_2), 3
PTMEAS/CART, 0,106,–36,–1,0,0
PTMEAS/CART, 0,79,–36,–1,0,0
PTMEAS/CART, 0,71,–21,–1,0,0

GOTO/–5,71,–21
GOTO/–5,71,20
GOTO/–5,4,20
GOTO/51,4,20
GOTO/51,4,20
ENDMES

F(StOneDp_3)=FEAT/PLANE, CART, 55,110,0,0,0,1
MEAS / PLANE, F(StOneDp_3), 3
PTMEAS/CART, 51,4,0,0,0,1
PTMEAS/CART, 51,106,0,0,0,1
PTMEAS/CART, 19,106,0,0,0,1

GOTO/19,106,20
GOTO/19,4,20
GOTO/6,4,20
GOTO/6,4,0
ENDMES

F(StOneDp_4)=FEAT/PLANE, CART, 10,110,–20,0,0,1
MEAS / PLANE, F(StOneDp_4), 3
```

```
PTMEAS/CART, 6,4,–20,0,0,1
PTMEAS/CART, 6,106,–20,0,0,1
PTMEAS/CART, 4,106,–20,0,0,1

GOTO/4,106,20
GOTO/4,106,20
GOTO/60,106,20
GOTO/60,106,–11
ENDMES

F(StMiddL_5)=FEAT/PLANE, CART, 55,110,0,1,0,0
MEAS / PLANE, F(StMiddL_5), 3
PTMEAS/CART, 55,106,–11,1,0,0
PTMEAS/CART, 55,4,–11,1,0,0
PTMEAS/CART, 55,4,–4,1,0,0

GOTO/60,4,–4
GOTO/60,4,20
GOTO/60,106,20
GOTO/80,106,20
GOTO/80,106,–11
ENDMES

F(StMiddL_6)=FEAT/PLANE, CART, 85,0,0,–1,0,0
MEAS / PLANE, F(StMiddL_6), 3
PTMEAS/CART, 85,106,–11,–1,0,0
PTMEAS/CART, 85,106,–4,–1,0,0
PTMEAS/CART, 85,4,–4,–1,0,0

GOTO/80,4,–4
GOTO/80,4,20
GOTO/80,4,20
GOTO/121,4,20
GOTO/121,4,20
ENDMES
```

```
F(StMidDp_7)=FEAT/PLANE, CART, 85,0,0,0,0,1
MEAS / PLANE, F(StMidDp_7), 3
PTMEAS/CART, 121,4,0,0,0,1
PTMEAS/CART, 121,106,0,0,0,1
PTMEAS/CART, 89,106,0,0,0,1

GOTO/89,106,20
GOTO/89,4,20
GOTO/76,4,20
GOTO/76,4,0
ENDMES

F(StMidDp_8)=FEAT/PLANE, CART, 60,0,–20,0,0,1
MEAS / PLANE, F(StMidDp_8), 3
PTMEAS/CART, 76,4,–20,0,0,1
PTMEAS/CART, 76,106,–20,0,0,1
PTMEAS/CART, 64,106,–20,0,0,1

GOTO/64,106,20
GOTO/64,4,20
GOTO/121,4,20
GOTO/121,4,20
ENDMES

F(StTwoDp_9)=FEAT/PLANE, CART, 85,0,0,0,0,1
MEAS / PLANE, F(StTwoDp_9), 3
PTMEAS/CART, 121,4,0,0,0,1
PTMEAS/CART, 121,106,0,0,0,1
PTMEAS/CART, 89,106,0,0,0,1

GOTO/89,106,20
GOTO/89,4,20
GOTO/136,4,20
GOTO/136,4,0
ENDMES
```

```
F(StTwoDp_10)=FEAT/PLANE, CART, 130,0,–20,0,0,1
MEAS / PLANE, F(StTwoDp_10), 3
PTMEAS/CART, 136,4,–20,0,0,1
PTMEAS/CART, 136,106,–20,0,0,1
PTMEAS/CART, 134,106,–20,0,0,1

GOTO/134,106,20
GOTO/134,106,20
GOTO/130,106,20
GOTO/130,106,–11
ENDMES

F(StTwoL_11)=FEAT/PLANE, CART, 125,0,–15,1,0,0
MEAS / PLANE, F(StTwoL_11), 3
PTMEAS/CART, 125,106,–11,1,0,0
PTMEAS/CART, 125,106,–4,1,0,0
PTMEAS/CART, 125,4,–4,1,0,0

GOTO/130,4,–4
GOTO/130,4,20
GOTO/130,106,20
GOTO/145,106,20
GOTO/145,106,–36
ENDMES

F(StTwoL_12)=FEAT/PLANE, CART, 140,110,–40,1,0,0
MEAS / PLANE, F(StTwoL_12), 3
PTMEAS/CART, 140,106,–36,1,0,0
PTMEAS/CART, 140,106,–24,1,0,0
PTMEAS/CART, 140,4,–24,1,0,0

GOTO/145,4,–24
GOTO/145,4,20
GOTO/145,106,20
GOTO/80,106,20
```

```
GOTO/80,106,–11
ENDMES

F(StMidRg_13)=FEAT/PLANE, CART, 85,0,0,–1,0,0
MEAS / PLANE, F(StMidRg_13), 3
PTMEAS/CART, 85,106,–11,–1,0,0
PTMEAS/CART, 85,106,–4,–1,0,0
PTMEAS/CART, 85,4,–4,–1,0,0

GOTO/80,4,–4
GOTO/80,4,20
GOTO/80,4,20
GOTO/76,4,20
GOTO/76,4,0
ENDMES

F(StMidBt_14)=FEAT/PLANE, CART, 60,0,–20,0,0,1
MEAS / PLANE, F(StMidBt_14), 3
PTMEAS/CART, 76,4,–20,0,0,1
PTMEAS/CART, 76,106,–20,0,0,1
PTMEAS/CART, 64,106,–20,0,0,1

GOTO/64,106,20
GOTO/64,106,20
GOTO/60,106,20
GOTO/60,106,–11
ENDMES

F(StMidLf_15)=FEAT/PLANE, CART, 55,110,0,1,0,0
MEAS / PLANE, F(StMidLf_15), 3
PTMEAS/CART, 55,106,–11,1,0,0
PTMEAS/CART, 55,4,–11,1,0,0
PTMEAS/CART, 55,4,–4,1,0,0

GOTO/60,4,–4
GOTO/60,4,20
```

```
GOTO/60,17.5,20
GOTO/27.5,17.5,20
GOTO/27.5,17.5,20
ENDMES

F(FrHole_16)=FEAT/CYLNDR, INNER, CART,
27.5,17.5,–20.0,0,0,1,20,–40
MEAS / CYLNDR, F(FrHole_16), 12
PTMEAS/CART, 37.5,17.5,–4,1,0,0
PTMEAS/CART, 27.5,27.5,–4,0,–1,0
PTMEAS/CART, 17.5,17.5,–4,–1,0,0
PTMEAS/CART, 27.5,7.5,–4,0,1,0
PTMEAS/CART, 37.5,17.5,–8,1,0,0
PTMEAS/CART, 27.5,27.5,–8,0,–1,0
PTMEAS/CART, 17.5,17.5,–8,–1,0,0
PTMEAS/CART, 27.5,7.5,–8,0,1,0
PTMEAS/CART, 37.5,17.5,–12,1,0,0
PTMEAS/CART, 27.5,27.5,–12,0,–1,0
PTMEAS/CART, 17.5,17.5,–12,–1,0,0
PTMEAS/CART, 27.5,7.5,–12,0,1,0

GOTO/27.5,17.5,20
GOTO/27.5,17.5,20
GOTO/112.5,17.5,20
GOTO/112.5,17.5,20
ENDMES

F(ScHole_17)=FEAT/CYLNDR, INNER, CART,
112.5,17.5,–20.0,0,0,1,20,–40
MEAS / CYLNDR, F(ScHole_17), 12
PTMEAS/CART, 122.5,17.5,–4,1,0,0
PTMEAS/CART, 112.5,27.5,–4,0,–1,0
PTMEAS/CART, 102.5,17.5,–4,–1,0,0
PTMEAS/CART, 112.5,7.5,–4,0,1,0
PTMEAS/CART, 122.5,17.5,–8,1,0,0
```

```
PTMEAS/CART, 112.5,27.5,–8,0,–1,0
PTMEAS/CART, 102.5,17.5,–8,–1,0,0
PTMEAS/CART, 112.5,7.5,–8,0,1,0
PTMEAS/CART, 122.5,17.5,–12,1,0,0
PTMEAS/CART, 112.5,27.5,–12,0,–1,0
PTMEAS/CART, 102.5,17.5,–12,–1,0,0
PTMEAS/CART, 112.5,7.5,–12,0,1,0

GOTO/112.5,17.5,20
GOTO/112.5,92.5,20
GOTO/27.5,92.5,20
GOTO/27.5,92.5,20
ENDMES

F(ThHole_18)=FEAT/CYLNDR, INNER, CART,
27.5,92.5,–20.0,0,0,1,20,–40
MEAS / CYLNDR, F(ThHole_18), 12
PTMEAS/CART, 37.5,92.5,–4,1,0,0
PTMEAS/CART, 27.5,102.5,–4,0,–1,0
PTMEAS/CART, 17.5,92.5,–4,–1,0,0
PTMEAS/CART, 27.5,82.5,–4,0,1,0
PTMEAS/CART, 37.5,92.5,–8,1,0,0
PTMEAS/CART, 27.5,102.5,–8,0,–1,0
PTMEAS/CART, 17.5,92.5,–8,–1,0,0
PTMEAS/CART, 27.5,82.5,–8,0,1,0
PTMEAS/CART, 37.5,92.5,–12,1,0,0
PTMEAS/CART, 27.5,102.5,–12,0,–1,0
PTMEAS/CART, 17.5,92.5,–12,–1,0,0
PTMEAS/CART, 27.5,82.5,–12,0,1,0

GOTO/27.5,92.5,20
GOTO/27.5,92.5,20
GOTO/112.5,92.5,20
GOTO/112.5,92.5,20
ENDMES
```

```
F(FuHole_19)=FEAT/CYLNDR, INNER, CART,
112.5,92.5,–20.0,0,0,1,20,–40
MEAS / CYLNDR, F(FuHole_19), 12
PTMEAS/CART, 122.5,92.5,–4,1,0,0
PTMEAS/CART, 112.5,102.5,–4,0,–1,0
PTMEAS/CART, 102.5,92.5,–4,–1,0,0
PTMEAS/CART, 112.5,82.5,–4,0,1,0
PTMEAS/CART, 122.5,92.5,–8,1,0,0
PTMEAS/CART, 112.5,102.5,–8,0,–1,0
PTMEAS/CART, 102.5,92.5,–8,–1,0,0
PTMEAS/CART, 112.5,82.5,–8,0,1,0
PTMEAS/CART, 122.5,92.5,–12,1,0,0
PTMEAS/CART, 112.5,102.5,–12,0,–1,0
PTMEAS/CART, 102.5,92.5,–12,–1,0,0
PTMEAS/CART, 112.5,82.5,–12,0,1,0

ENDMES

T(genera_1)=TOL/DISTB , NOMINL,15, 0.05, 0.05, PT2PT
OUTPUT/FA(StOneL_1),FA(StOneL_2), TA(genera_1)

T(genera_3)=TOL/DISTB , NOMINL,20, 0.05, 0.05, PT2PT
OUTPUT/FA(StOneDp_3),FA(StOneDp_4), TA(genera_3)

T(genera_5)=TOL/DISTB , NOMINL,30, 0.05, 0.05, PT2PT
OUTPUT/FA(StMiddL_5),FA(StMiddL_6), TA(genera_5)

T(genera_7)=TOL/DISTB , NOMINL, 20, 0.05, 0.05, PT2PT
OUTPUT/FA(StMidDp_7),FA(StMidDp_8), TA(genera_7)

T(genera_9)=TOL/DISTB , NOMINL, 20, 0.05, 0.05, PT2PT
OUTPUT/FA(StTwoDp_9),FA(StTwoDp_10), TA(genera_9)

T(genera_11)=TOL/DISTB , NOMINL, 15, 0.05, 0.05, PT2PT
OUTPUT/FA(StTwoL_11),FA(StTwoL_12), TA(genera_11)

T(flatne_13)=TOL/FLAT, 0.05
OUTPUT/FA(StMidRg_13), TA(flatne_13)
```

```
T(parall_14)=TOL/PARLEL, 0.001
OUTPUT/FA(StMidBt_14), TA(parall_14)

T(perpen_15)=TOL/PERP, 0.001,FA(AA_0)
OUTPUT/FA(StMidLf_15), TA(perpen_15)

T(cylind_16)=TOL/CYLCTY, 0.001
OUTPUT/FA(FrHole_16), TA(cylind_16)

T(circul_17)=TOL/CIRLTY, 0.001
OUTPUT/FA(ScHole_17), TA(circul_17)

T(circul_18)=TOL/CIRLTY, 0.001
OUTPUT/FA(ThHole_18), TA(circul_18)

T(cylind_19)=TOL/CYLCTY, 0.001
OUTPUT/FA(FuHole_19), TA(cylind_19)

ENDFIL
```

A1.3 DMIS Code Programming of Third Case Study

```
$$ CARL ZEISS – CALYPSO Preprocessor
$$ Ver.1.01.029.00 Date: Sat Dec 29 2012 Time: 12:24:51

DMISMN/'Case Study No. 3'
FILNAM/'CaseThree.DMI'
UNITS/MM, ANGDEC, TEMPC

SNSET/APPRCH, 2.0000
SNSET/RETRCT, 5
SNSET/SEARCH, 5.0000
```

```
$$ Changing label 'Star ~ 1' to: 'STAR1'.
S(STAR1)=SNSDEF/PROBE, INDEX, CART, –0.0150, 0.1975,
–20.9113, 0.0000, 0.0000, $
–1.0000, 3.0050, SPHERE

RECALL/DA(REFERENCE)
MODE/PROG,MAN

F(AA_0)=FEAT/PLANE, CART, 140,75,0,0,0,1
MEAS / PLANE, F(AA_0), 4
PTMEAS/CART, 133,82,0,0,0,1
PTMEAS/CART, 133,103,0,0,0,1
PTMEAS/CART, 7,103,0,0,0,1
PTMEAS/CART, 7,68,0,0,0,1

GOTO/7,68,20
GOTO/7,42,20
GOTO/133,42,20
GOTO/133,42,10
ENDMES

F(SlMidDp_1)=FEAT/PLANE, CART, 0,35,–10,0,0,1
MEAS / PLANE, F(SlMidDp_1), 4
PTMEAS/CART, 133,42,–10,0,0,1
PTMEAS/CART, 133,68,–10,0,0,1
PTMEAS/CART, 7,68,–10,0,0,1
PTMEAS/CART, 7,28,–10,0,0,1

GOTO/7,28,20
GOTO/7,82,20
GOTO/133,82,20
GOTO/133,82,20
ENDMES

F(SlMidDp_2)=FEAT/PLANE, CART, 140,75,0,0,0,1
MEAS / PLANE, F(SlMidDp_2), 4
```

```
PTMEAS/CART, 133,82,0,0,0,1
PTMEAS/CART, 133,103,0,0,0,1
PTMEAS/CART, 7,103,0,0,0,1
PTMEAS/CART, 7,68,0,0,0,1

GOTO/7,68,20
GOTO/7,117,20
GOTO/145,117,20
GOTO/145,117,–23
ENDMES

F(SlMidLg_3)=FEAT/PLANE, CART, 140,110,–30,1,0,0
MEAS / PLANE, F(SlMidLg_3), 4
PTMEAS/CART, 140,117,–23,1,0,0
PTMEAS/CART, 140,103,7,1,0,0
PTMEAS/CART, 140,68,–7,1,0,0
PTMEAS/CART, 140,68,–3,1,0,0

GOTO/145,68,–3
GOTO/145,68,20
GOTO/145,117,20
GOTO/–5,117,20
GOTO/–5,117,–7
ENDMES

F(SlMidLg_4)=FEAT/PLANE, CART, 0,110,0,–1,0,0
MEAS / PLANE, F(SlMidLg_4), 4
PTMEAS/CART, 0,117,–7,–1,0,0
PTMEAS/CART, 0,103,–37,–1,0,0
PTMEAS/CART, 0,97,–37,–1,0,0
PTMEAS/CART, 0,83,–47,–1,0,0

GOTO/–5,83,–47
GOTO/–5,83,20
GOTO/–5,17.5,20
```

```
GOTO/35,17.5,20
GOTO/35,17.5,20
ENDMES

F(FrHolCy_5)=FEAT/CYLNDR, INNER, CART,
35.,17.5,–5.0,0,0,1,20,–10
MEAS / CYLNDR, F(FrHolCy_5), 16
PTMEAS/CART, 45,17.5,–1,1,0,0
PTMEAS/CART, 35,27.5,–1,0,–1,0
PTMEAS/CART, 25,17.5,–1,–1,0,0
PTMEAS/CART, 35,7.5,–1,0,1,0
PTMEAS/CART, 45,17.5,–2,1,0,0
PTMEAS/CART, 35,27.5,–2,0,–1,0
PTMEAS/CART, 25,17.5,–2,–1,0,0
PTMEAS/CART, 35,7.5,–2,0,1,0
PTMEAS/CART, 45,17.5,–3,1,0,0
PTMEAS/CART, 35,27.5,–3,0,–1,0
PTMEAS/CART, 25,17.5,–3,–1,0,0
PTMEAS/CART, 35,7.5,–3,0,1,0
PTMEAS/CART, 45,17.5,–4,1,0,0
PTMEAS/CART, 35,27.5,–4,0,–1,0
PTMEAS/CART, 25,17.5,–4,–1,0,0
PTMEAS/CART, 35,7.5,–4,0,1,0

GOTO/35,17.5,20
GOTO/35,17.5,20
GOTO/35,17.5,20
GOTO/35,17.5,20
ENDMES

F(FrHolCr_6)=FEAT/CYLNDR, INNER, CART,
35.,17.5,–5.0,0,0,1,20,–10
MEAS / CYLNDR, F(FrHolCr_6), 16
PTMEAS/CART, 45,17.5,–1,1,0,0
PTMEAS/CART, 35,27.5,–1,0,–1,0
```

```
PTMEAS/CART, 25,17.5,–1,–1,0,0
PTMEAS/CART, 35,7.5,–1,0,1,0
PTMEAS/CART, 45,17.5,–2,1,0,0
PTMEAS/CART, 35,27.5,–2,0,–1,0
PTMEAS/CART, 25,17.5,–2,–1,0,0
PTMEAS/CART, 35,7.5,–2,0,1,0
PTMEAS/CART, 45,17.5,–3,1,0,0
PTMEAS/CART, 35,27.5,–3,0,–1,0
PTMEAS/CART, 25,17.5,–3,–1,0,0
PTMEAS/CART, 35,7.5,–3,0,1,0
PTMEAS/CART, 45,17.5,–4,1,0,0
PTMEAS/CART, 35,27.5,–4,0,–1,0
PTMEAS/CART, 25,17.5,–4,–1,0,0
PTMEAS/CART, 35,7.5,–4,0,1,0

GOTO/35,17.5,20
GOTO/35,17.5,20
GOTO/105,17.5,20
GOTO/105,17.5,20
ENDMES

F(ScHolCy_7)=FEAT/CYLNDR, INNER, CART,
105.,17.5,–5.0,0,0,1,20,–10
MEAS / CYLNDR, F(ScHolCy_7), 16
PTMEAS/CART, 115,17.5,–1,1,0,0
PTMEAS/CART, 105,27.5,–1,0,–1,0
PTMEAS/CART, 95,17.5,–1,–1,0,0
PTMEAS/CART, 105,7.5,–1,0,1,0
PTMEAS/CART, 115,17.5,–2,1,0,0
PTMEAS/CART, 105,27.5,–2,0,–1,0
PTMEAS/CART, 95,17.5,–2,–1,0,0
PTMEAS/CART, 105,7.5,–2,0,1,0
PTMEAS/CART, 115,17.5,–3,1,0,0
PTMEAS/CART, 105,27.5,–3,0,–1,0
```

```
PTMEAS/CART, 95,17.5,–3,–1,0,0
PTMEAS/CART, 105,7.5,–3,0,1,0
PTMEAS/CART, 115,17.5,–4,1,0,0
PTMEAS/CART, 105,27.5,–4,0,–1,0
PTMEAS/CART, 95,17.5,–4,–1,0,0
PTMEAS/CART, 105,7.5,–4,0,1,0

GOTO/105,17.5,20
GOTO/105,92.5,20
GOTO/35,92.5,20
GOTO/35,92.5,20
ENDMES

F(ThHolCy_8)=FEAT/CYLNDR, INNER, CART,
35.,92.5,–5.0,0,0,1,20,–10
MEAS / CYLNDR, F(ThHolCy_8), 16
PTMEAS/CART, 45,92.5,–1,1,0,0
PTMEAS/CART, 35,102.5,–1,0,–1,0
PTMEAS/CART, 25,92.5,–1,–1,0,0
PTMEAS/CART, 35,82.5,–1,0,1,0
PTMEAS/CART, 45,92.5,–2,1,0,0
PTMEAS/CART, 35,102.5,–2,0,–1,0
PTMEAS/CART, 25,92.5,–2,–1,0,0
PTMEAS/CART, 35,82.5,–2,0,1,0
PTMEAS/CART, 45,92.5,–3,1,0,0
PTMEAS/CART, 35,102.5,–3,0,–1,0
PTMEAS/CART, 25,92.5,–3,–1,0,0
PTMEAS/CART, 35,82.5,–3,0,1,0
PTMEAS/CART, 45,92.5,–4,1,0,0
PTMEAS/CART, 35,102.5,–4,0,–1,0
PTMEAS/CART, 25,92.5,–4,–1,0,0
PTMEAS/CART, 35,82.5,–4,0,1,0

GOTO/35,92.5,20
GOTO/35,92.5,20
```

```
GOTO/105,92.5,20
GOTO/105,92.5,20
ENDMES

F(FuHolCy_9)=FEAT/CYLNDR, INNER, CART,
105.,92.5,–5.0,0,0,1,20,–10
MEAS / CYLNDR, F(FuHolCy_9), 16
PTMEAS/CART, 115,92.5,–1,1,0,0
PTMEAS/CART, 105,102.5,–1,0,–1,0
PTMEAS/CART, 95,92.5,–1,–1,0,0
PTMEAS/CART, 105,82.5,–1,0,1,0
PTMEAS/CART, 115,92.5,–2,1,0,0
PTMEAS/CART, 105,102.5,–2,0,–1,0
PTMEAS/CART, 95,92.5,–2,–1,0,0
PTMEAS/CART, 105,82.5,–2,0,1,0
PTMEAS/CART, 115,92.5,–3,1,0,0
PTMEAS/CART, 105,102.5,–3,0,–1,0
PTMEAS/CART, 95,92.5,–3,–1,0,0
PTMEAS/CART, 105,82.5,–3,0,1,0
PTMEAS/CART, 115,92.5,–4,1,0,0
PTMEAS/CART, 105,102.5,–4,0,–1,0
PTMEAS/CART, 95,92.5,–4,–1,0,0
PTMEAS/CART, 105,82.5,–4,0,1,0

GOTO/105,92.5,20
GOTO/133,92.5,20
GOTO/133,70,20
GOTO/133,70,–3
ENDMES

F(SlMidFr_10)=FEAT/PLANE, CART, 140,75,0,0,–1,0
MEAS / PLANE, F(SlMidFr_10), 4
PTMEAS/CART, 133,75,–3,0,–1,0
PTMEAS/CART, 7,75,–3,0,–1,0
```

```
PTMEAS/CART, 7,75,–7,0,–1,0
PTMEAS/CART, 133,75,7,0,–1,0

GOTO/133,70,7
GOTO/133,70,20
GOTO/133,70,20
GOTO/133,42,20
GOTO/133,42,10
ENDMES

F(SlMidBo_11)=FEAT/PLANE, CART, 0,35,–10,0,0,1
MEAS / PLANE, F(SlMidBo_11), 4
PTMEAS/CART, 133,42,–10,0,0,1
PTMEAS/CART, 133,68,–10,0,0,1
PTMEAS/CART, 7,68,–10,0,0,1
PTMEAS/CART, 7,28,–10,0,0,1

ENDMES

T(genera_1)=TOL/DISTB , NOMINL, 10,–0.05, 0.05,PT2PT
OUTPUT/FA(SlMidDp_1) FA(SlMidDp_2), TA(genera_1)

T(genera_3)=TOL/DISTB , NOMINL, 140,–0.05, 0.05,PT2PT
OUTPUT/FA(SlMidLg_3) FA(SlMidLg_4), TA(genera_3)

T(cylind_5)=TOL/CYLCTY, 0.001
OUTPUT/FA(FrHolCy_5), TA(cylind_5)

T(circul_6)=TOL/CIRLTY, 0.001
OUTPUT/FA(FrHolCr_6), TA(circul_6)

T(cylind_7)=TOL/CYLCTY, 0.001
OUTPUT/FA(ScHolCy_7), TA(cylind_7)

T(cylind_8)=TOL/CYLCTY, 0.001
OUTPUT/FA(ThHolCy_8), TA(cylind_8)

T(cylind_9)=TOL/CYLCTY, 0.001
OUTPUT/FA(FuHolCy_9), TA(cylind_9)
```

```
T(perpen_10)=TOL/PERP, 0.05
OUTPUT/FA(SlMidFr_10), TA(perpen_10)

T(flatne_11)=TOL/FLAT, 0.002,FA( _AA_11)
OUTPUT/FA(SlMidBo_11), TA(flatne_11)

ENDFIL
```

A1.4 DMIS Code Programming of Hub at the First Setup

```
$$ CARL ZEISS – CALYPSO Preprocessor
$$ Ver.1.01.029.00 Date: Thu Dec 27 2012 Time: 22:03:13

DMISMN/'FIRSTPLAN'
FILNAM/'FIRSTPLAN.DMI'
UNITS/MM, ANGDEC, TEMPC

SNSET/APPRCH, 2.0000
SNSET/RETRCT, 5
SNSET/SEARCH, 5.0000

$$ Changing label 'Star ~ 1' to: 'STAR1'.
S(STAR1)=SNSDEF/PROBE, INDEX, CART, –0.0150, 0.1975,
–20.9113, 0.0000, 0.0000, $ –1.0000, 3.0050, SPHERE

RECALL/DA(REFERENCE)

MODE/PROG,MAN

F(FrHole_0)=FEAT/CYLNDR, INNER, CART, –63.4999999999,
–107.95,–30.16250000010,0,0,1,12,–15.875
MEAS / CYLNDR, F(FrHole_0), 16
PTMEAS/CART, –57.5,–107.95,–1.5875,1,0,0
```

```
PTMEAS/CART, –63.5,–101.95,–1.5875,0,–1,0
PTMEAS/CART, –69.5,–107.95,–1.5875,–1,0,0
PTMEAS/CART, –63.5,–113.95,–1.5875,0,1,0
PTMEAS/CART, –57.5,–107.95,–3.175,1,0,0
PTMEAS/CART, –63.5,–101.95,–3.175,0,–1,0
PTMEAS/CART, –69.5,–107.95,–3.175,–1,0,0
PTMEAS/CART, –63.5,–113.95,–3.175,0,1,0
PTMEAS/CART, –57.5,–107.95,–4.7625,1,0,0
PTMEAS/CART, –63.5,–101.95,–4.7625,0,–1,0
PTMEAS/CART, –69.5,–107.95,–4.7625,–1,0,0
PTMEAS/CART, –63.5,–113.95,–4.7625,0,1,0
PTMEAS/CART, –57.5,–107.95,–6.35,1,0,0
PTMEAS/CART, –63.5,–101.95,–6.35,0,–1,0
PTMEAS/CART, –69.5,–107.95,–6.35,–1,0,0
PTMEAS/CART, –63.5,–113.95,–6.35,0,1,0

GOTO/–63.5,–107.95,20
GOTO/–63.5,–63.5,20
GOTO/–107.95,–63.5,20
GOTO/–107.95,–63.5,20
ENDMES

F(ScHole_1)=FEAT/CYLNDR, INNER, CART, –107.95,–63.5,
–30.16250000010,0,0,1,12,–15.875
MEAS / CYLNDR, F(ScHole_1), 16
PTMEAS/CART, –101.95,–63.5,–1.5875,1,0,0
PTMEAS/CART, –107.95,–57.5,–1.5875,0,–1,0
PTMEAS/CART, –113.95,–63.5,–1.5875,–1,0,0
PTMEAS/CART, –107.95,–69.5,–1.5875,0,1,0
PTMEAS/CART, –101.95,–63.5,–3.175,1,0,0
PTMEAS/CART, –107.95,–57.5,–3.175,0,–1,0
PTMEAS/CART, –113.95,–63.5,–3.175,–1,0,0
PTMEAS/CART, –107.95,–69.5,–3.175,0,1,0
PTMEAS/CART, –101.95,–63.5,–4.7625,1,0,0
PTMEAS/CART, –107.95,–57.5,–4.7625,0,–1,0
```

```
PTMEAS/CART, –113.95,–63.5,–4.7625,–1,0,0
PTMEAS/CART, –107.95,–69.5,–4.7625,0,1,0
PTMEAS/CART, –101.95,–63.5,–6.35,1,0,0
PTMEAS/CART, –107.95,–57.5,–6.35,0,–1,0
PTMEAS/CART, –113.95,–63.5,–6.35,–1,0,0
PTMEAS/CART, –107.95,–69.5,–6.35,0,1,0

GOTO/–107.95,–63.5,20
GOTO/–107.95,–19.05,20
GOTO/–63.5,–19.05,20
GOTO/–63.5,–19.05,20
ENDMES

F(TrHole_2)=FEAT/CYLNDR, INNER, CART, –63.4999999999,
–19.0499999998,–30.16250000010,0,0,1,12,–15.875
MEAS / CYLNDR, F(TrHole_2), 16
PTMEAS/CART, –57.5,–19.05,–1.5875,1,0,0
PTMEAS/CART, –63.5,–13.05,–1.5875,0,–1,0
PTMEAS/CART, –69.5,–19.05,–1.5875,–1,0,0
PTMEAS/CART, –63.5,–25.05,–1.5875,0,1,0
PTMEAS/CART, –57.5,–19.05,–3.175,1,0,0
PTMEAS/CART, –63.5,–13.05,–3.175,0,–1,0
PTMEAS/CART, –69.5,–19.05,–3.175,–1,0,0
PTMEAS/CART, –63.5,–25.05,–3.175,0,1,0
PTMEAS/CART, –57.5,–19.05,–4.7625,1,0,0
PTMEAS/CART, –63.5,–13.05,–4.7625,0,–1,0
PTMEAS/CART, –69.5,–19.05,–4.7625,–1,0,0
PTMEAS/CART, –63.5,–25.05,–4.7625,0,1,0
PTMEAS/CART, –57.5,–19.05,–6.35,1,0,0
PTMEAS/CART, –63.5,–13.05,–6.35,0,–1,0
PTMEAS/CART, –69.5,–19.05,–6.35,–1,0,0
PTMEAS/CART, –63.5,–25.05,–6.35,0,1,0

GOTO/–63.5,–19.05,20
GOTO/–63.5,–63.5,20
```

```
GOTO/–19.05,–63.5,20
GOTO/–19.05,–63.5,20
ENDMES

F(FuHole_3)=FEAT/CYLNDR, INNER, CART, –19.0499999997,
–63.5,–30.16250000010,0,0,1,12,–15.875
MEAS / CYLNDR, F(FuHole_3), 16
PTMEAS/CART, –13.05,–63.5,–1.5875,1,0,0
PTMEAS/CART, –19.05,–57.5,–1.5875,0,–1,0
PTMEAS/CART, –25.05,–63.5,–1.5875,–1,0,0
PTMEAS/CART, –19.05,–69.5,–1.5875,0,1,0
PTMEAS/CART, –13.05,–63.5,–3.175,1,0,0
PTMEAS/CART, –19.05,–57.5,–3.175,0,–1,0
PTMEAS/CART, –25.05,–63.5,–3.175,–1,0,0
PTMEAS/CART, –19.05,–69.5,–3.175,0,1,0
PTMEAS/CART, –13.05,–63.5,–4.7625,1,0,0
PTMEAS/CART, –19.05,–57.5,–4.7625,0,–1,0
PTMEAS/CART, –25.05,–63.5,–4.7625,–1,0,0
PTMEAS/CART, –19.05,–69.5,–4.7625,0,1,0
PTMEAS/CART, –13.05,–63.5,–6.35,1,0,0
PTMEAS/CART, –19.05,–57.5,–6.35,0,–1,0
PTMEAS/CART, –25.05,–63.5,–6.35,–1,0,0
PTMEAS/CART, –19.05,–69.5,–6.35,0,1,0

GOTO/–19.05,–63.5,20
GOTO/–19.05,–63.5,20
GOTO/–63.5,–63.5,20
GOTO/–63.5,–63.5,20
ENDMES

F(UMHole_4)=FEAT/CYLNDR, INNER, CART,
–63.4999999999,–63.5,–3.175000000010,0,0,1,30,–6.35
MEAS / CYLNDR, F(UMHole_4), 16
PTMEAS/CART, –48.5,–63.5,–0.635,1,0,0
PTMEAS/CART, –63.5,–48.5,–0.635,0,–1,0
```

```
PTMEAS/CART, –78.5,–63.5,–0.635,–1,0,0
PTMEAS/CART, –63.5,–78.5,–0.635,0,1,0
PTMEAS/CART, –48.5,–63.5,–1.27,1,0,0
PTMEAS/CART, –63.5,–48.5,–1.27,0,–1,0
PTMEAS/CART, –78.5,–63.5,–1.27,–1,0,0
PTMEAS/CART, –63.5,–78.5,–1.27,0,1,0
PTMEAS/CART, –48.5,–63.5,–1.905,1,0,0
PTMEAS/CART, –63.5,–48.5,–1.905,0,–1,0
PTMEAS/CART, –78.5,–63.5,–1.905,–1,0,0
PTMEAS/CART, –63.5,–78.5,–1.905,0,1,0
PTMEAS/CART, –48.5,–63.5,–2.54,1,0,0
PTMEAS/CART, –63.5,–48.5,–2.54,0,–1,0
PTMEAS/CART, –78.5,–63.5,–2.54,–1,0,0
PTMEAS/CART, –63.5,–78.5,–2.54,0,1,0

GOTO/–63.5,–63.5,20
GOTO/–63.5,–63.5,20
GOTO/–63.5,–63.5,20
GOTO/–63.5,–63.5,20
ENDMES

F(DMHole_5)=FEAT/CYLNDR, INNER, CART,
–63.4999999999,–63.5,–22.22500000010,0,0,1,30,–19
MEAS / CYLNDR, F(DMHole_5), 16
PTMEAS/CART, –48.5,–63.5,–1.9,1,0,0
PTMEAS/CART, –63.5,–48.5,–1.9,0,–1,0
PTMEAS/CART, –78.5,–63.5,–1.9,–1,0,0
PTMEAS/CART, –63.5,–78.5,–1.9,0,1,0
PTMEAS/CART, –48.5,–63.5,–3.8,1,0,0
PTMEAS/CART, –63.5,–48.5,–3.8,0,–1,0
PTMEAS/CART, –78.5,–63.5,–3.8,–1,0,0
PTMEAS/CART, –63.5,–78.5,–3.8,0,1,0
PTMEAS/CART, –48.5,–63.5,–5.7,1,0,0
PTMEAS/CART, –63.5,–48.5,–5.7,0,–1,0
PTMEAS/CART, –78.5,–63.5,–5.7,–1,0,0
```

```
PTMEAS/CART, –63.5,–78.5,–5.7,0,1,0
PTMEAS/CART, –48.5,–63.5,–7.6,1,0,0
PTMEAS/CART, –63.5,–48.5,–7.6,0,–1,0
PTMEAS/CART, –78.5,–63.5,–7.6,–1,0,0
PTMEAS/CART, –63.5,–78.5,–7.6,0,1,0

GOTO/–63.5,–63.5,20
GOTO/–63.5,–56,20
GOTO/–122,–56,20
GOTO/–122,–56,–29
ENDMES

F(HUBWdth_6)=FEAT/PLANE, CART, –127,–75,–34,1,0,0
MEAS / PLANE, F(HUBWdth_6), 4
PTMEAS/CART, –127,–56,–29,1,0,0
PTMEAS/CART, –127,–70,–29,1,0,0
PTMEAS/CART, –127,–70,–24,1,0,0
PTMEAS/CART, –127,–61,–24,1,0,0

GOTO/–122,–61,–24
GOTO/–122,–61,20
GOTO/–122,–56,20
GOTO/–2,–56,20
GOTO/–2,–56,–29
ENDMES

F(HUBWdth_7)=FEAT/PLANE, CART, 3,–51,–34,–1,0,0
MEAS / PLANE, F(HUBWdth_7), 4
PTMEAS/CART, 3,–56,–29,–1,0,0
PTMEAS/CART, 3,–56,–24,–1,0,0
PTMEAS/CART, 3,–70,–24,–1,0,0
PTMEAS/CART, 3,–61,–24,–1,0,0

GOTO/–2,–61,–24
GOTO/–2,–61,20
```

```
GOTO/–58,–61,20
GOTO/–58,–132,20
GOTO/–58,–132,–29
ENDMES

F(HUBLnth_8)=FEAT/PLANE, CART, –53,–127,–34,0,–1,0
MEAS / PLANE, F(HUBLnth_8), 4
PTMEAS/CART, –58,–127,–29,0,–1,0
PTMEAS/CART, –58,–127,–24,0,–1,0
PTMEAS/CART, –68,–127,–24,0,–1,0
PTMEAS/CART, –68,–127,–29,0,–1,0

GOTO/–68,–132,–29
GOTO/–68,–132,20
GOTO/–58,–132,20
GOTO/–58,7,20
GOTO/–58,7,–29
ENDMES

F(HUBLnth_9)=FEAT/PLANE, CART, –73,2,–34,0,1,0
MEAS / PLANE, F(HUBLnth_9), 4
PTMEAS/CART, –58,2,–29,0,1,0
PTMEAS/CART, –68,2,–29,0,1,0
PTMEAS/CART, –68,2,–24,0,1,0
PTMEAS/CART, –58,2,–24,0,1,0

GOTO/–58,7,–24
GOTO/–58,7,20
GOTO/–15,7,20
GOTO/–15,–90,20
GOTO/–15,–90,–20
ENDMES

F(FuLenth_10)=FEAT/PLANE, CART, –31,–85,–25,0,–1,0
MEAS / PLANE, F(FuLenth_10), 4
```

```
PTMEAS/CART, –15,–85,–20,0,–1,0
PTMEAS/CART, –26,–85,–20,0,–1,0
PTMEAS/CART, –26,–85,–24,0,–1,0
PTMEAS/CART, –15,–85,–24,0,–1,0

GOTO/–15,–90,–24
GOTO/–15,–90,20
GOTO/–15,–90,20
GOTO/–15,–36,20
GOTO/–15,–36,–20
ENDMES

F(FuLenth_11)=FEAT/PLANE, CART, –10,–41,–25,0,1,0
MEAS / PLANE, F(FuLenth_11), 4
PTMEAS/CART, –15,–41,–20,0,1,0
PTMEAS/CART, –15,–41,–24,0,1,0
PTMEAS/CART, –26,–41,–24,0,1,0
PTMEAS/CART, –26,–41,–20,0,1,0

GOTO/–26,–36,–20
GOTO/–26,–36,20
GOTO/–26,–17,20
GOTO/–80,–17,20
GOTO/–80,–17,–20
ENDMES

F(TrLenth_12)=FEAT/PLANE, CART, –85,–12,–25,1,0,0
MEAS / PLANE, F(TrLenth_12), 4
PTMEAS/CART, –85,–17,–20,1,0,0
PTMEAS/CART, –85,–17,–24,1,0,0
PTMEAS/CART, –85,–26,–24,1,0,0
PTMEAS/CART, –85,–12,–25,1,0,0

GOTO/–80,–12,–25
GOTO/–80,–12,20
```

```
GOTO/–80,–17,20
GOTO/–46,–17,20
GOTO/–46,–17,–20
ENDMES

F(TrLenth_13)=FEAT/PLANE, CART, –41,–31,–25,–1,0,0
MEAS / PLANE, F(TrLenth_13), 4
PTMEAS/CART, –41,–17,–20,–1,0,0
PTMEAS/CART, –41,–26,–20,–1,0,0
PTMEAS/CART, –41,–26,–24,–1,0,0
PTMEAS/CART, –41,–22,–15,–1,0,0

GOTO/–46,–22,–15
GOTO/–46,–22,20
GOTO/–100,–22,20
GOTO/–100,–90,20
GOTO/–100,–90,–20
ENDMES

F(ScLenth_14)=FEAT/PLANE, CART, –116,–85,–25,0,–1,0
MEAS / PLANE, F(ScLenth_14), 4
PTMEAS/CART, –100,–85,–20,0,–1,0
PTMEAS/CART, –111,–85,–20,0,–1,0
PTMEAS/CART, –111,–85,–24,0,–1,0
PTMEAS/CART, –100,–85,–24,0,–1,0

GOTO/–100,–90,–24
GOTO/–100,–90,20
GOTO/–100,–90,20
GOTO/–100,–36,20
GOTO/–100,–36,–20
ENDMES

F(ScLenth_15)=FEAT/PLANE, CART, –95,–41,–25,0,1,0
MEAS / PLANE, F(ScLenth_15), 4
```

```
PTMEAS/CART, –100,–41,–20,0,1,0
PTMEAS/CART, –100,–41,–24,0,1,0
PTMEAS/CART, –111,–41,–24,0,1,0
PTMEAS/CART, –111,–41,–20,0,1,0

GOTO/–111,–36,–20
GOTO/–111,–36,20
GOTO/–111,–100,20
GOTO/–46,–100,20
GOTO/–46,–100,–20
ENDMES

F(FrLenth_16)=FEAT/PLANE, CART, –41,–114,–25,–1,0,0
MEAS / PLANE, F(FrLenth_16), 4
PTMEAS/CART, –41,–100,–20,–1,0,0
PTMEAS/CART, –41,–109,–20,–1,0,0
PTMEAS/CART, –41,–109,–24,–1,0,0
PTMEAS/CART, –41,–105,–15,–1,0,0

GOTO/–46,–105,–15
GOTO/–46,–105,20
GOTO/–46,–100,20
GOTO/–80,–100,20
GOTO/–80,–100,–20
ENDMES

F(FrLenth_17)=FEAT/PLANE, CART, –85,–95,–25,1,0,0
MEAS / PLANE, F(FrLenth_17), 4
PTMEAS/CART, –85,–100,–20,1,0,0
PTMEAS/CART, –85,–100,–24,1,0,0
PTMEAS/CART, –85,–109,–24,1,0,0
PTMEAS/CART, –85,–95,–25,1,0,0

GOTO/–80,–95,–25
GOTO/–80,–95,20
```

```
GOTO/−80,−107.95,20
GOTO/−63.5,−107.95,20
GOTO/−63.5,−107.95,20
ENDMES

T(cylind_0)=TOL/CYLCTY, 0.005
OUTPUT/FA(FrHole_0), TA(cylind_0)

T(cylind_1)=TOL/CYLCTY, 0.005
OUTPUT/FA(ScHole_1), TA(cylind_1)

T(cylind_2)=TOL/CYLCTY, 0.005
OUTPUT/FA(TrHole_2), TA(cylind_2)

T(cylind_3)=TOL/CYLCTY, 0.005
OUTPUT/FA(FuHole_3), TA(cylind_3)

T(cylind_4)=TOL/CYLCTY, 0.004
OUTPUT/FA(UMHole_4), TA(cylind_4)

T(cylind_5)=TOL/CYLCTY, 0.004
OUTPUT/FA(DMHole_5), TA(cylind_5)

T(genera_6)=TOL/DISTB, NOMINL, 130,−0.05, 0.05,PT2PT
OUTPUT/FA(HUBWdth_6) FA(HUBWdth_7), TA(genera_6)

T(genera_8)=TOL/DISTB, NOMINL, 129,−0.05, 0.05,PT2PT
OUTPUT/FA(HUBLnth_8) FA(HUBLnth_9), TA(genera_8)

T(genera_10)=TOL/DISTB, NOMINL, 44,−0.05, 0.05,PT2PT
OUTPUT/FA(FuLenth_10) FA(FuLenth_11), TA(genera_10)

T(genera_12)=TOL/DISTB, NOMINL, 44,−0.05, 0.05,PT2PT
OUTPUT/FA(TrLenth_12) FA(TrLenth_13), TA(genera_12)
```

```
T(genera_14)=TOL/DISTB, NOMINL, 44,–0.05, 0.05,PT2PT
OUTPUT/FA(ScLenth_14) FA(ScLenth_15), TA(genera_14)

T(genera_16)=TOL/DISTB, NOMINL, 44,–0.05, 0.05,PT2PT
OUTPUT/FA(FrLenth_16) FA(FrLenth_17), TA(genera_16)

ENDFIL
```

A1.5 DMIS Code Programming of Hub at the Second Setup

```
$$ CARL ZEISS – CALYPSO Preprocessor
$$ Ver.1.01.029.00 Date: Thu Dec 27 2012 Time: 22:05:52

DMISMN/'REFERENCE'
FILNAM/'REFERENCE.DMI'
UNITS/MM, ANGDEC, TEMPC

SNSET/APPRCH, 2.0000
SNSET/RETRCT, 5
SNSET/SEARCH, 5.0000

$$ Changing label 'Star ~ 1' to: 'STAR1'.
S(STAR1)=SNSDEF/PROBE, INDEX, CART, –0.0150, 0.1975,
–20.9113, 0.0000, 0.0000, $
–1.0000, 3.0050, SPHERE

RECALL/DA(REFERENCE)

MODE/PROG,MAN

F(FrHole_0)=FEAT/CYLNDR, INNER, CART, –18.4999999999,
–20.,–3.324999999730,0,0,1,10,–10.16
MEAS / CYLNDR, F(FrHole_0), 16
```

```
PTMEAS/CART, –13.5,–20,–1.016,1,0,0
PTMEAS/CART, –18.5,–15,–1.016,0,–1,0
PTMEAS/CART, –23.5,–20,–1.016,–1,0,0
PTMEAS/CART, –18.5,–25,–1.016,0,1,0
PTMEAS/CART, –13.5,–20,–2.032,1,0,0
PTMEAS/CART, –18.5,–15,–2.032,0,–1,0
PTMEAS/CART, –23.5,–20,–2.032,–1,0,0
PTMEAS/CART, –18.5,–25,–2.032,0,1,0
PTMEAS/CART, –13.5,–20,–3.048,1,0,0
PTMEAS/CART, –18.5,–15,–3.048,0,–1,0
PTMEAS/CART, –23.5,–20,–3.048,–1,0,0
PTMEAS/CART, –18.5,–25,–3.048,0,1,0
PTMEAS/CART, –13.5,–20,–4.064,1,0,0
PTMEAS/CART, –18.5,–15,–4.064,0,–1,0
PTMEAS/CART, –23.5,–20,–4.064,–1,0,0
PTMEAS/CART, –18.5,–25,–4.064,0,1,0

GOTO/–18.5,–20,20
GOTO/–18.5,–107,20
GOTO/–18.5,–107,20
GOTO/–18.5,–107,20
ENDMES

F(ScHole_1)=FEAT/CYLNDR, INNER, CART, –18.49999999
99,–107.,–3.324999999730,0,0,1,10,–10.16
MEAS / CYLNDR, F(ScHole_1), 16
PTMEAS/CART, –13.5,–107,–1.016,1,0,0
PTMEAS/CART, –18.5,–102,–1.016,0,–1,0
PTMEAS/CART, –23.5,–107,–1.016,–1,0,0
PTMEAS/CART, –18.5,–112,–1.016,0,1,0
PTMEAS/CART, –13.5,–107,–2.032,1,0,0
PTMEAS/CART, –18.5,–102,–2.032,0,–1,0
PTMEAS/CART, –23.5,–107,–2.032,–1,0,0
PTMEAS/CART, –18.5,–112,–2.032,0,1,0
```

```
PTMEAS/CART, –13.5,–107,–3.048,1,0,0
PTMEAS/CART, –18.5,–102,–3.048,0,–1,0
PTMEAS/CART, –23.5,–107,–3.048,–1,0,0
PTMEAS/CART, –18.5,–112,–3.048,0,1,0
PTMEAS/CART, –13.5,–107,–4.064,1,0,0
PTMEAS/CART, –18.5,–102,–4.064,0,–1,0
PTMEAS/CART, –23.5,–107,–4.064,–1,0,0
PTMEAS/CART, –18.5,–112,–4.064,0,1,0

GOTO/–18.5,–107,20
GOTO/–18.5,–107,20
GOTO/–108.5,–107,20
GOTO/–108.5,–107,20
ENDMES

F(ThHole_2)=FEAT/CYLNDR, INNER, CART, –108.5,–107.,
–3.324999999730,0,0,1,10,–10.16
MEAS / CYLNDR, F(ThHole_2), 16
PTMEAS/CART, –103.5,–107,–1.016,1,0,0
PTMEAS/CART, –108.5,–102,–1.016,0,–1,0
PTMEAS/CART, –113.5,–107,–1.016,–1,0,0
PTMEAS/CART, –108.5,–112,–1.016,0,1,0
PTMEAS/CART, –103.5,–107,–2.032,1,0,0
PTMEAS/CART, –108.5,–102,–2.032,0,–1,0
PTMEAS/CART, –113.5,–107,–2.032,–1,0,0
PTMEAS/CART, –108.5,–112,–2.032,0,1,0
PTMEAS/CART, –103.5,–107,–3.048,1,0,0
PTMEAS/CART, –108.5,–102,–3.048,0,–1,0
PTMEAS/CART, –113.5,–107,–3.048,–1,0,0
PTMEAS/CART, –108.5,–112,–3.048,0,1,0
PTMEAS/CART, –103.5,–107,–4.064,1,0,0
PTMEAS/CART, –108.5,–102,–4.064,0,–1,0
PTMEAS/CART, –113.5,–107,–4.064,–1,0,0
PTMEAS/CART, –108.5,–112,–4.064,0,1,0
```

```
GOTO/–108.5,–107,20
GOTO/–108.5,–20,20
GOTO/–108.5,–20,20
GOTO/–108.5,–20,20
ENDMES

F(FuHole_3)=FEAT/CYLNDR, INNER, CART, –108.5,–20.,
–3.324999999730,0,0,1,10,–10.16
MEAS / CYLNDR, F(FuHole_3), 16
PTMEAS/CART, –103.5,–20,–1.016,1,0,0
PTMEAS/CART, –108.5,–15,–1.016,0,–1,0
PTMEAS/CART, –113.5,–20,–1.016,–1,0,0
PTMEAS/CART, –108.5,–25,–1.016,0,1,0
PTMEAS/CART, –103.5,–20,–2.032,1,0,0
PTMEAS/CART, –108.5,–15,–2.032,0,–1,0
PTMEAS/CART, –113.5,–20,–2.032,–1,0,0
PTMEAS/CART, –108.5,–25,–2.032,0,1,0
PTMEAS/CART, –103.5,–20,–3.048,1,0,0
PTMEAS/CART, –108.5,–15,–3.048,0,–1,0
PTMEAS/CART, –113.5,–20,–3.048,–1,0,0
PTMEAS/CART, –108.5,–25,–3.048,0,1,0
PTMEAS/CART, –103.5,–20,–4.064,1,0,0
PTMEAS/CART, –108.5,–15,–4.064,0,–1,0
PTMEAS/CART, –113.5,–20,–4.064,–1,0,0
PTMEAS/CART, –108.5,–25,–4.064,0,1,0

GOTO/–108.5,–20,20
GOTO/–36,–20,20
GOTO/–36,–56,20
GOTO/–36,–56,–30
ENDMES

F(LnthBot_4)=FEAT/PLANE, CART, –41,–51,–35,0,–1,0
MEAS / PLANE, F(LnthBot_4), 4
PTMEAS/CART, –36,–51,–30,0,–1,0
```

```
PTMEAS/CART, –36,–51,–30,0,–1,0
PTMEAS/CART, –36,–51,–11,0,–1,0
PTMEAS/CART, –36,–51,–11,0,–1,0

GOTO/–36,–56,–11
GOTO/–36,–56,20
GOTO/–36,–56,20
GOTO/–36,–70,20
GOTO/–36,–70,–30
ENDMES

F(LnthBot_5)=FEAT/PLANE, CART, –31,–75,–6,0,1,0
MEAS / PLANE, F(LnthBot_5), 4
PTMEAS/CART, –36,–75,–30,0,1,0
PTMEAS/CART, –36,–75,–30,0,1,0
PTMEAS/CART, –36,–75,–11,0,1,0
PTMEAS/CART, –36,–75,–11,0,1,0

GOTO/–36,–70,–11
GOTO/–36,–70,20
GOTO/–36,–92,20
GOTO/–58,–92,20
GOTO/–58,–92,–30
ENDMES

F(WdthLft_6)=FEAT/PLANE, CART, –53,–87,–35,–1,0,0
MEAS / PLANE, F(WdthLft_6), 4
PTMEAS/CART, –53,–92,–30,–1,0,0
PTMEAS/CART, –53,–92,–11,–1,0,0
PTMEAS/CART, –53,–90,–11,–1,0,0
PTMEAS/CART, –53,–97,–25,–1,0,0

GOTO/–58,–97,–25
GOTO/–58,–97,20
GOTO/–58,–92,20
```

```
GOTO/–68,–92,20
GOTO/–68,–92,–30
ENDMES

F(WdthLft_7)=FEAT/PLANE, CART, –73,–95,–6,1,0,0
MEAS / PLANE, F(WdthLft_7), 4
PTMEAS/CART, –73,–92,–30,1,0,0
PTMEAS/CART, –73,–92,–11,1,0,0
PTMEAS/CART, –73,–90,–11,1,0,0
PTMEAS/CART, –73,–97,–25,1,0,0

GOTO/–68,–97,–25
GOTO/–68,–97,20
GOTO/–90,–97,20
GOTO/–90,–70,20
GOTO/–90,–70,–30
ENDMES

F(LnthUpr_8)=FEAT/PLANE, CART, –85,–75,–35,0,1,0
MEAS / PLANE, F(LnthUpr_8), 4
PTMEAS/CART, –90,–75,–30,0,1,0
PTMEAS/CART, –90,–75,–11,0,1,0
PTMEAS/CART, –90,–75,–11,0,1,0
PTMEAS/CART, –90,–75,–30,0,1,0

GOTO/–90,–70,–30
GOTO/–90,–70,20
GOTO/–90,–70,20
GOTO/–90,–56,20
GOTO/–90,–56,–30
ENDMES

F(LnthUpr_9)=FEAT/PLANE, CART, –95,–51,–6,0,–1,0
MEAS / PLANE, F(LnthUpr_9), 4
PTMEAS/CART, –90,–51,–30,0,–1,0
```

```
PTMEAS/CART, –90,–51,–11,0,–1,0
PTMEAS/CART, –90,–51,–11,0,–1,0
PTMEAS/CART, –90,–51,–30,0,–1,0

GOTO/–90,–56,–30
GOTO/–90,–56,20
GOTO/–78,–56,20
GOTO/–78,–36,20
GOTO/–78,–36,–30
ENDMES

F(WdthRit_10)=FEAT/PLANE, CART, –73,–39,–35,–1,0,0
MEAS / PLANE, F(WdthRit_10), 4
PTMEAS/CART, –73,–36,–30,–1,0,0
PTMEAS/CART, –73,–34,–30,–1,0,0
PTMEAS/CART, –73,–34,–11,–1,0,0
PTMEAS/CART, –73,–31,–35,–1,0,0

GOTO/–78,–31,–35
GOTO/–78,–31,20
GOTO/–78,–36,20
GOTO/–48,–36,20
GOTO/–48,–36,–30
ENDMES

F(WdthRit_11)=FEAT/PLANE, CART, –53,–31,–6,1,0,0
MEAS / PLANE, F(WdthRit_11), 4
PTMEAS/CART, –53,–36,–30,1,0,0
PTMEAS/CART, –53,–34,–30,1,0,0
PTMEAS/CART, –53,–34,–11,1,0,0
PTMEAS/CART, –53,–31,–35,1,0,0

GOTO/–48,–31,–35
GOTO/–48,–31,20
```

```
GOTO/−48,−63.5,20
GOTO/−63.5,−63.5,20
GOTO/−63.5,−63.5,20
ENDMES

F(RoundCR_12)=FEAT/CYLNDR, INNER, CART, −63.49999
99999,−63.5,−39.83749999990,0,0,1,50,−28.575
MEAS / CYLNDR, F(RoundCR_12), 16
PTMEAS/CART, −38.5,−63.5,−2.8575,1,0,0
PTMEAS/CART, −63.5,−38.5,−2.8575,0,−1,0
PTMEAS/CART, −88.5,−63.5,−2.8575,−1,0,0
PTMEAS/CART, −63.5,−88.5,−2.8575,0,1,0
PTMEAS/CART, −38.5,−63.5,−5.715,1,0,0
PTMEAS/CART, −63.5,−38.5,−5.715,0,−1,0
PTMEAS/CART, −88.5,−63.5,−5.715,−1,0,0
PTMEAS/CART, −63.5,−88.5,−5.715,0,1,0
PTMEAS/CART, −38.5,−63.5,−8.5725,1,0,0
PTMEAS/CART, −63.5,−38.5,−8.5725,0,−1,0
PTMEAS/CART, −88.5,−63.5,−8.5725,−1,0,0
PTMEAS/CART, −63.5,−88.5,−8.5725,0,1,0
PTMEAS/CART, −38.5,−63.5,−11.43,1,0,0
PTMEAS/CART, −63.5,−38.5,−11.43,0,−1,0
PTMEAS/CART, −88.5,−63.5,−11.43,−1,0,0
PTMEAS/CART, −63.5,−88.5,−11.43,0,1,0

ENDMES

T(cylind_0)=TOL/CYLCTY, 0.005
OUTPUT/FA(FrHole_0), TA(cylind_0)

T(cylind_1)=TOL/CYLCTY, 0.005
OUTPUT/FA(ScHole_1), TA(cylind_1)

T(cylind_2)=TOL/CYLCTY, 0.005
OUTPUT/FA(ThHole_2), TA(cylind_2)
```

```
T(cylind_3)=TOL/CYLCTY, 0.005
OUTPUT/FA(FuHole_3), TA(cylind_3)

T(genera_4)=TOL/DISTB , NOMINL, 24,–0.05, 0.05,PT2PT
OUTPUT/FA(LnthBot_4) FA(LnthBot_5), TA(genera_4)

T(genera_6)=TOL/DISTB , NOMINL, 20,–0.05, 0.05,PT2PT
OUTPUT/FA(WdthLft_6) FA(WdthLft_7), TA(genera_6)

T(genera_8)=TOL/DISTB , NOMINL, 24,–0.05, 0.05,PT2PT
OUTPUT/FA(LnthUpr_8) FA(LnthUpr_9), TA(genera_8)

T(genera_10)=TOL/DISTB , NOMINL, 20,–0.05, 0.05,PT2PT
OUTPUT/FA(WdthRit_10) FA(WdthRit_11), TA(genera_10)

T(cylind_12)=TOL/CYLCTY, 0.005
OUTPUT/FA(RoundCR_12), TA(cylind_12)

ENDFIL
```

A1.6 DMIS Code Programming of Gear Pump Housing at the First Setup

```
$$ CARL ZEISS – CALYPSO Preprocessor
$$ Ver.1.01.029.00 Date: Tue Apr 9 2013 Time: 13:55:58

DMISMN/'SSSSSSSSS'

FILNAM/'SSSSSSSSS.DMI'

UNITS/MM, ANGDEC, TEMPC

SNSET/APPRCH, 2.0000
SNSET/RETRCT, 5
SNSET/SEARCH, 5.0000
```

```
$$ Changing label 'Star ~ 1' to: 'STAR1'.
S(STAR1)=SNSDEF/PROBE,INDEX,CART,–0.0150,
0.1975,–20.9113, 0.0000, 0.0000, $
–1.0000, 3.0050, SPHERE

RECALL/DA(REFERENCE)

MODE/PROG,MAN

F(FrHoleD_1)=FEAT/CYLNDR, INNER, CART,,
0,77.4,00,0,–1,0,14,32
MEAS / CYLNDR, F(FrHoleD_1), 8
PTMEAS/CART, 7,48.7,0,1,0,0
PTMEAS/CART, 0,48.7,7,0,0,1
PTMEAS/CART, –7,48.7,0,–1,0,0
PTMEAS/CART, 0,48.7,–7,0,0,–1
PTMEAS/CART, 7,51.9,0,1,0,0
PTMEAS/CART, 0,51.9,7,0,0,1
PTMEAS/CART, –7,51.9,0,–1,0,0
PTMEAS/CART, 0,51.9,–7,0,0,–1

GOTO/0,–5,77.4
GOTO/0,–5,20
GOTO/0,–5,20
GOTO/0,–5,20
GOTO/0,–5,77.4
ENDMES

F(ScHoleD_2)=FEAT/CYLNDR, INNER, CART,,
0,77.4,15.90,0,–1,0,14,32
MEAS / CYLNDR, F(ScHoleD_2), 8
PTMEAS/CART, 7,48.7,15.9,1,0,0
PTMEAS/CART, 0,48.7,22.9,0,0,1
PTMEAS/CART, –7,48.7,15.9,–1,0,0
```

```
PTMEAS/CART, 0,48.7,8.9,0,0,–1
PTMEAS/CART, 7,51.9,15.9,1,0,0
PTMEAS/CART, 0,51.9,22.9,0,0,1
PTMEAS/CART, –7,51.9,15.9,–1,0,0
PTMEAS/CART, 0,51.9,8.9,0,0,–1

GOTO/0,–5,77.4
GOTO/0,–5,20
GOTO/0,–5,20
GOTO/0,–5,20
GOTO/0,–5,77.4
ENDMES

F(ThHoleD_3)=FEAT/CYLNDR, INNER, CART,,
33.35,77.4,00,0,–1,0,14,32
MEAS / CYLNDR, F(ThHoleD_3), 8
PTMEAS/CART, 40.35,48.7,0,1,0,0
PTMEAS/CART, 33.35,48.7,7,0,0,1
PTMEAS/CART, 26.35,48.7,0,–1,0,0
PTMEAS/CART, 33.35,48.7,–7,0,0,–1
PTMEAS/CART, 40.35,51.9,0,1,0,0
PTMEAS/CART, 33.35,51.9,7,0,0,1
PTMEAS/CART, 26.35,51.9,0,–1,0,0
PTMEAS/CART, 33.35,51.9,–7,0,0,–1

GOTO/0,–5,77.4
GOTO/0,–5,20
GOTO/0,–5,20
GOTO/0,–5,20
GOTO/0,–5,77.4
ENDMES

F(FuHoleD_4)=FEAT/CYLNDR, INNER, CART,,
33.35,77.4,15.90,0,–1,0,14,32
MEAS / CYLNDR, F(FuHoleD_4), 8
```

```
PTMEAS/CART, 40.35,48.7,15.9,1,0,0
PTMEAS/CART, 33.35,48.7,22.9,0,0,1
PTMEAS/CART, 26.35,48.7,15.9,–1,0,0
PTMEAS/CART, 33.35,48.7,8.9,0,0,–1
PTMEAS/CART, 40.35,51.9,15.9,1,0,0
PTMEAS/CART, 33.35,51.9,22.9,0,0,1
PTMEAS/CART, 26.35,51.9,15.9,–1,0,0
PTMEAS/CART, 33.35,51.9,8.9,0,0,–1

GOTO/0,–5,77.4
GOTO/0,–5,20
GOTO/5,–5,20
GOTO/5,28.225,20
GOTO/5,28.225,0

ENDMES

T(cylind_0)=TOL/CYLCTY, 0.005
OUTPUT/FA(RightCy_0), TA(cylind_0)

T(cylind_1)=TOL/CYLCTY, 0.005
OUTPUT/FA(FrHoleD_1), TA(cylind_1)

T(cylind_2)=TOL/CYLCTY, 0.005
OUTPUT/FA(ScHoleD_2), TA(cylind_2)

T(cylind_3)=TOL/CYLCTY, 0.005
OUTPUT/FA(ThHoleD_3), TA(cylind_3)

T(cylind_4)=TOL/CYLCTY, 0.005
OUTPUT/FA(FuHoleD_4), TA(cylind_4)

ENDFIL
```

A1.7 DMIS Code Programming of Gear Pump Housing at the Second Setup

```
$$ CARL ZEISS – CALYPSO Preprocessor
$$ Ver.1.01.029.00 Date: Tue Apr 9 2013 Time: 13:59:58

DMISMN/'SAM'

FILNAM/'SAM.DMI'

UNITS/MM, ANGDEC, TEMPC

SNSET/APPRCH, 2.0000
SNSET/RETRCT, 5
SNSET/SEARCH, 5.0000

$$ Changing label 'Star ~ 1' to: 'STAR1'.
S(STAR1)=SNSDEF/PROBE, INDEX, CART, –0.0150, 0.1975,
–20.9113, 0.0000, 0.0000, $
–1.0000, 3.0050, SPHERE

RECALL/DA(REFERENCE)

MODE/PROG,MAN

F(LfFHolD_5)=FEAT/CYLNDR,INNER,CART,
28.225,0,45.0,1,0,0,16,–112.9
MEAS / CYLNDR, F(LfFHolD_5), 8
PTMEAS/CART, 45.16,8,45,0,1,0
PTMEAS/CART, 45.16,0,53,0,0,1
PTMEAS/CART, 45.16,–8,45,0,–1,0
PTMEAS/CART, 45.16,0,37,0,0,1
PTMEAS/CART, 33.87,8,45,0,1,0
PTMEAS/CART, 33.87,0,53,0,0,1
PTMEAS/CART, 33.87,–8,45,0,–1,0
PTMEAS/CART, 33.87,0,37,0,0,1
```

```
GOTO/5,28.225,0
GOTO/5,28.225,20
GOTO/5,28.225,20
GOTO/–5,28.225,20
GOTO/–5,28.225,55
ENDMES

F(RgFHolD_6)=FEAT/CYLNDR,INNER,CART,,
28.225,55.,45.0,–,0,0,16,112.9
MEAS / CYLNDR, F(RgFHolD_6), 8
PTMEAS/CART, –45.16,63,45,0,1,0
PTMEAS/CART, –45.16,55,53,0,0,1
PTMEAS/CART, –45.16,47,45,0,–1,0
PTMEAS/CART, –45.16,55,37,0,0,–1
PTMEAS/CART, –33.87,63,45,0,1,0
PTMEAS/CART, –33.87,55,53,0,0,1
PTMEAS/CART, –33.87,47,45,0,–1,0
PTMEAS/CART, –33.87,55,37,0,0,–1

ENDMES

T(cylind_5)=TOL/CYLCTY, 0.004
OUTPUT/FA(LfFHolD_5), TA(cylind_5)

T(cylind_6)=TOL/CYLCTY, 0.004
OUTPUT/FA(RgFHolD_6), TA(cylind_6)

ENDFIL
```

Index

A